Yvo JACQUIER

THE BOOK OF
GEOMETRY
WITH THE EYES

Basis of Composition in the Arts

INTRODUCTION

The purpose of this book is to explain in a didactic way an unprecedented chapter of geometry, whose elements are rooted in pre-Euclidean traditions. We highlight the mathematical coherence of this corpus, anchored on the notion of similarity to which the name of Thales is often attributed.

The term "geometry with the eyes" emphasizes the visual character of the (de) demonstrations, and this goes hand in hand with its oral transmission. This approach considerably reduces the necessity of the calculation of which we have the reflex. This makes it possible to envisage a very distant origin, long before the advent of writing.

This denomination is well adapted since these mathematics served the visual arts – painting and architecture -- until the Renaissance. Finally, this qualification is hitherto unusual. Its use, with a definition, should not create any confusion.

Research as a whole

The studies that preceded this publication are an example of trans-disciplinarity. For nearly fifteen years, several aspects have progressed in concert, and each field has respected its own methodology. The strictly mathematical part can be extracted from this set, and it has been realized in partnership with mathematicians of the IREM - in particular Jean-Paul Guichard of the IREM of Poitiers. This pure theoretical part is not only the guarantor of the whole, it is its heart. All the debates met there.

As new as it is, geometry with the eyes is not a creation in the strict sense, a new theory as that mathematicians can produce. It is the result of a "reading" applied to works of art and architecture, in one of the most misunderstood aspects: composition. The identification of this geometry has over time become a discipline in its own right.

The Comparative Geometry deploys a set of innovative methods such as margins of precision, systematic comparison of works and redoubling of observation from their geometry. These principles come together under the term methodology.

Composition is an intelligent art. The symbolism it carries is not merely narrative, and it is not enough just a few attributes to translate it. Its principles are mathematical and metaphysical. Jean-Paul Guichard tells us that for François Viète, the word "symbol" in mathematics has the meaning of incontestable truth. But in the visual arts, the study shows that the symbolism is borne by geometrical properties, which are demonstrated. They are based on the theoretical part described in this book.

These structures must find a meaning in the historical reality of the works, before confronting the decisive stage of their interpretation. The key to trans-disciplinarity is to offer to fertile hypotheses the possibility of broadening their field of definition; It allows them to reach a level of arguments and credibility that are difficult to envisage inside a single discipline. The composition is thus not only a question of mathematics, or of history, or of symbolic meaning. All these aspects combine together originally in the minds of the creators and ultimately in the mind of the reader.

Throughout this book, we will not neglect totally the different aspects of the initial overall research, but we will concentrate on a single subject: geometry with the eyes.

The most surprising discovery of this work concerns the four manifestations of the golden ratio in the triangle 3-4-5. The divine proportion is not presented here as a "strange phenomenon". It is one of the principles among others that make it possible to construct figures of composition.

NB : *The PDF version of the book offers color visuals. They can be accessed at : http://www.art-renaissance.net/G. The large size of the typography of this book makes it easy to handle it while displaying the images on the screen.*

I – THE ELEMENTS OF MONSTRATION

To do mathematics, we need definitions, logic and axioms. The definitions we need are very basic. The Sumerians and the Egyptians conceived them in a simple way. The Greeks, especially Euclid, will specify these notions. Note that the logic of the geometry we are approaching is very already exact. It is based, in particular, on relevant axioms – these things one admits.

1 – Thinking with angles

Before measuring shapes with instruments (knotted rope, ruler or compass), man first compared the shapes, particularly their angles, using his eye as the first tool of "judgment".

Fig. 1 – The diagonal of similar triangles

According to Jean-Paul Guichard, quoting Choquet (1964) "The notion of angle is undoubtedly the one that raises the most discussions and difficulties in the teaching of Geometry.

The angle, as it appears implicitly in our minds today, is an invention of the Greeks! All the known expressions that precede its completion, pedagogical and conceptual, are translated concretely by measurement of physical parameters. (see V, 1) (see bibliography)

« Mesopotamian and Egyptian astronomers measured the height of the stars and celestial bodies on the horizon. And it is always thus that astronomers or navigators express themselves. The Egyptian architects spoke of the slope of the pyramids (the *sekhed*), the Mesopotamian architects from that of their walls (the *fruit*). And this is always how roofers, masons or road signs are expressed. »

Thus the measurement of angle, the first consecrated notion of geometry with the eyes, was not "conceivable" in our contemporary terms by the societies that developed the knowledge we are approaching. They thought of it on a grid, and they relied on its tiles to talk about it - "oral tradition oblige". However, the spirit of similarity constantly finds in the angle its first visual argument. One would have to turn to the neurosciences, in terms of the ability to recognize forms, to conclude.

We face several aspects at once, each one becoming concern. 1 - The plausible restitution of a practice that would have developed in Mesopotamia and Egypt - especially. 2 - The (re) construction of a grid geometry that is fully coherent in modern terms, although pre-Euclidean. 3 - All of that, with didactic and even pedagogical will in these digitalized times ... It seems difficult to assume these three preoccupations at the same time, unless we keep three discourses permanently, three languages; not to mention the visuals that would need to be adapted. We will have to choose a sort of barycenter that allows common sense to form an idea, even if some specialists chose to take the virtuous distance of study. We will thus resort to shortcuts, sprains of language and notation, with the main result of making this work accessible.

2 – Symmetry, the first logical notion

Fig. 2 : The circle is the perfect example of symmetry. The first approach to geometry is concerned with angles, and it compares forms - particularly angles. Applied to the circle, this approach produces an observation: The angle that a straight line which touches the circle with the diameter that passes at this point is equal to each side of this diameter. This observation is "obvious" on a grid.

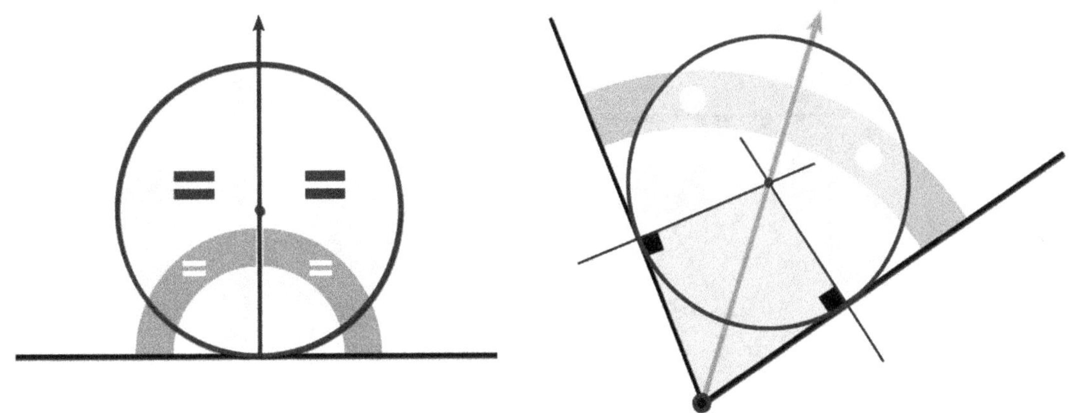

Fig. 3 : Particular example of symmetry: the bisector of any angle passes through the center of any circle which touches the two lines of this angle. The two halves of the circle shared by the bisector are symmetrical. It follows the appearance of a kite that will be one of the basic figures of all the demonstrations of geometry with the eyes.

3 – The inscribed circle of the triangle

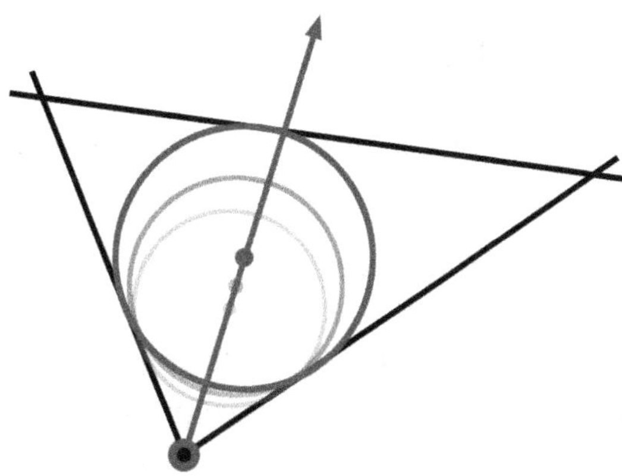

Fig. 4 : Inscribed circle in an any triangle

The existence of one and only one circle inscribed in the triangle then proceeds from common sense. It is enough to enlarge the circle which grows on the bisector of an angle until touching the third side. And the three bisectors can only cross at the center of this circle inscribed.

The first demonstrations did not need to go so far in their extension, for the grid reduces the figures to particular cases. The approach we give here is, in a way, that of the Greek mathematicians who had learned from Egypt.

4 – The principle of rotation

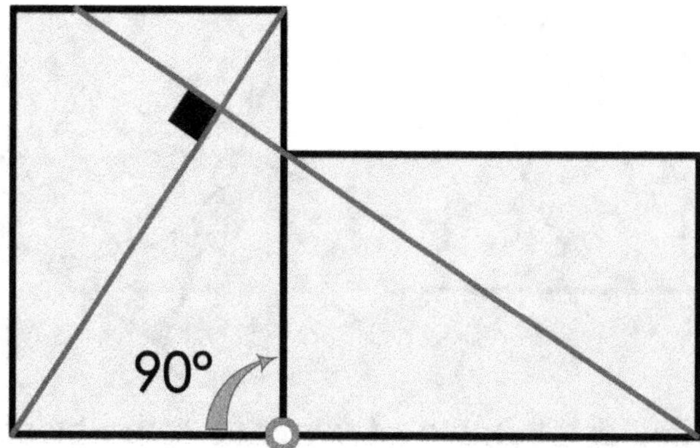

Fig 5 : Principle of rotation of a geometric shape.

Another element of demonstration: By the rotation of a quarter turn (90°), all the elements are found at right angle - as here with any rectangle. The diagonal of the rectangle produced by the rotation is at right angles to the diagonal of the initial rectangle (half flat angle).

5 – The sum of the angles of the triangle

It is sufficient, by translation, to join three identical triangles according to their three different angles at the same point O to show it: The sum of the angles of a triangle is 180°. (Or π, flat angle, or half-turn).

Fig 6 : The angle of two straight lines is the same on both sides of the point where they intersect, in particular the angle alpha. This demonstration is of the Euclidean type, but it is admissible and useful in a pedagogical framework.

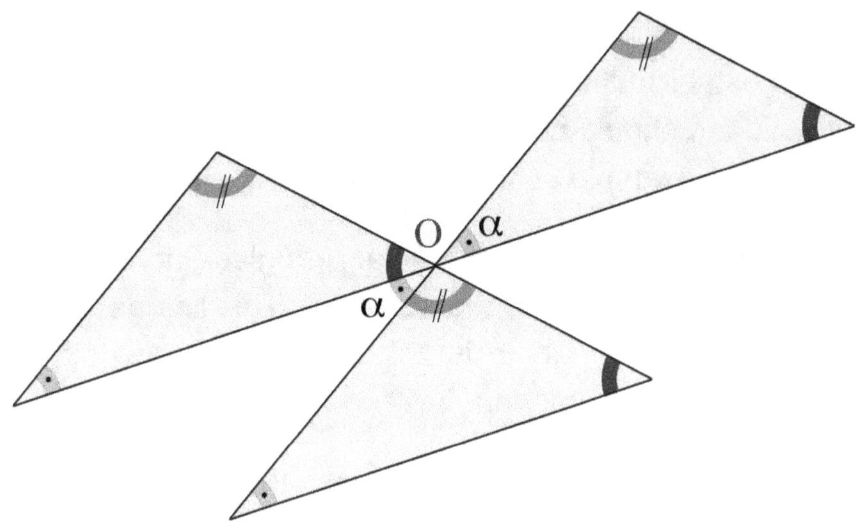

Fig 6 : Sum of the angles of the triangle

Fig 7 : Jean-Paul Guichard remarks that this is even more simple: Olivier Keller reports a series of Paleolithic friezes, engraved on a support in bone. Their structure allows the same observation.

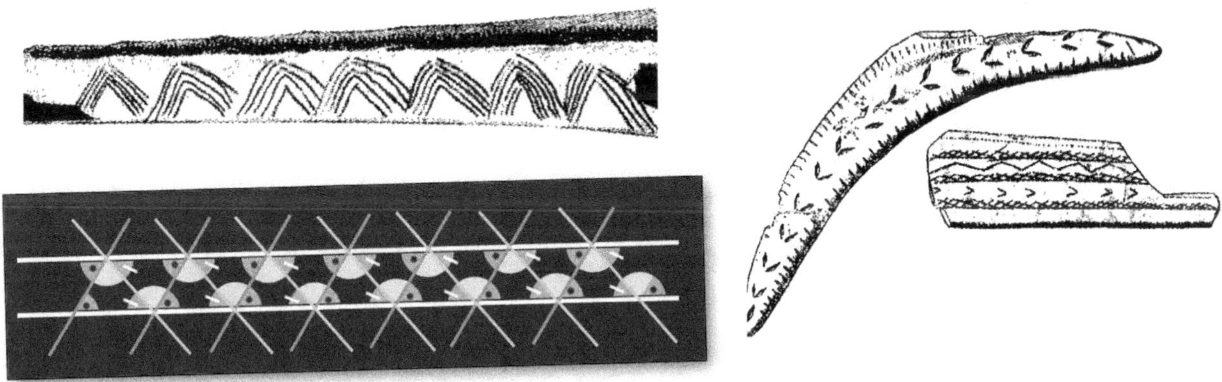

Fig 7 : Visuals taken from the book by Olivier Keller
« La figure et le monde, une archéologie de la géométrie »

6 – Similar triangles

Two triangles are said to be similar if   (and only if) their three angles are equal.

The sum of the angles of a triangle is a flat angle. In the comparison of two triangles, it is sufficient to note the equality of two of their angles to establish their similarity. The third angle proceeds from the same subtraction. It should be noted that this observation is made initially "with the eyes", for example considering the previous frieze.

In the particular case of rectangular triangles, the identity of one of the two angles of the hypotenuse is sufficient. This link between the non-right angles of the right triangle is at the origin of a whole chapter of mathematics, especially Sumerian...

A question arises: when has been the notion of equality of proportions revealed? For the comparison of the triangles, and for the comparison of the sides of the two triangles, taken separately. When exactly was this knowledge, that Thales would establish "officially", born?

With some common sense, we can reconstruct the stages of this conquest for the spirit. As we have pointed out, the first is that of angles, the first approach of stone-cutters and archers of the Upper Paleolithic.

Fig. 8 : Logically, the first triangle of the study is rectangle, placed on a straight grid. The particular cases (below on the left, the proportion 3 on 2) alert men in search of numbers (counting) and measurement (distance). The angle thus becomes a ratio between two measurements. This will become trigonometry.

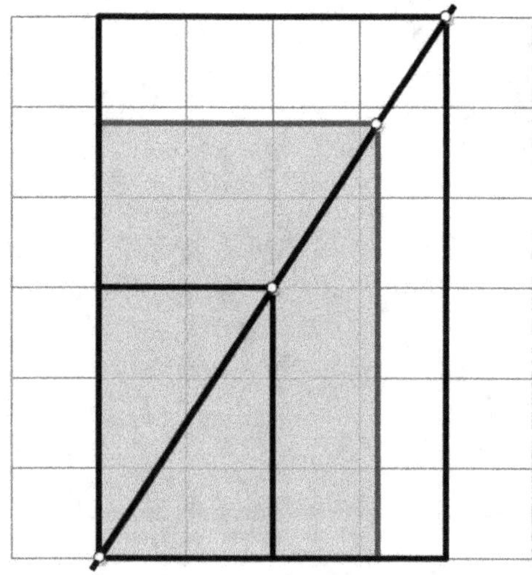

Fig. 8 : right-angled triangle

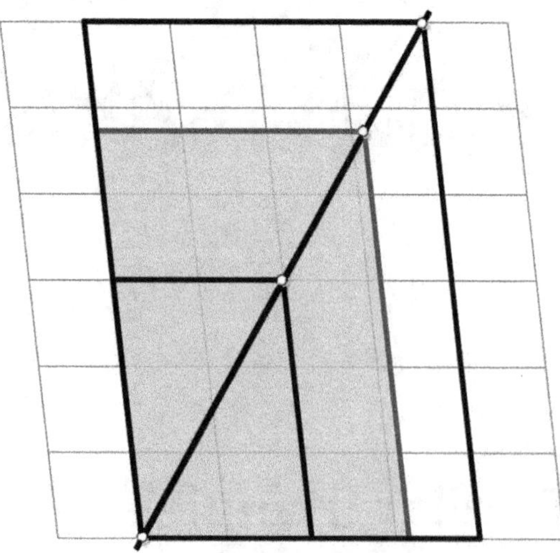

Fig. 9 : Any triangle

A first generalization will extend this observation about the integers, with respect to simple triangles, to a broader axiomatic rule :

The similar right-angled triangles have the same proportions.

Fig. 9 : Second stage of generalization: the extension of the axiom to any triangle. The same figure inclines the lines of its grid, which retains the same unit of measurement. The same type of reasoning leads to a rule, generalization of the preceding one :

Similar triangles have the same proportions.

7 – Particular case of the double-square

After we will find a particular case of similar triangles. They have the same right angle and their right-angled sides have the same proportion (= 2).

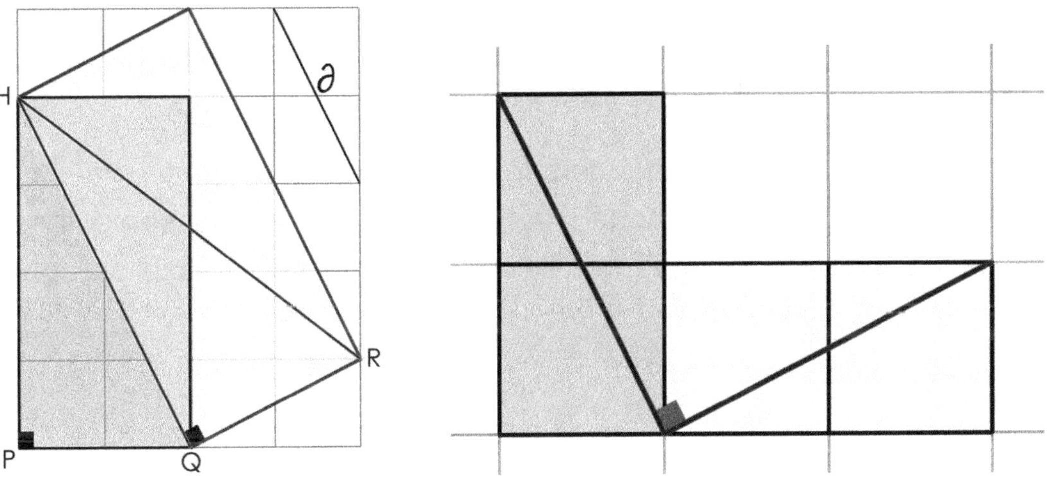

Fig. 10 a) and b) : Double-square

The demonstration begins with this observation : HQ is orthogonal to QR, because QR is obtained by rotating a double square of the grid (detailed here on the right). The triangle 2-1-∂ is called the Pinwheel triangle, and we will find it again.

Then HQ and HR are the diagonals of two similar rectangles, with a height / width ratio equal to 2.

The right-angled triangle is often considered as half a rectangle in the demonstrations within geometry with the eyes.

8 – Concepts that go ahead

The concept of area

The corpus of geometry with the eyes regularly comes into contact with the notion of area, but its status has logically been constructed in stages, from the principle of puzzle to calculation. *For those interested in this particular subject, see the bibliography.*

Similarity

On the same way, the idea of similarity has evolved since the recognition of a resemblance (for example that of similar triangles) to the Euclidean concept of transforming distances according to a constant.

9 – The presence of instruments

This culture seeks the certainty of evidences, what are called axioms and properties in modern terms, and it is at the same time a practice which does not break with its instruments. (see the bibliography)

II – FIGURES OF QUADRILLAGE

1 – The Grid

The grid is the frame of any reflection for geometry with the eyes. It allows the skillful bypassing of the calculation, and even avoiding it. The numbers are there. Integers, fractions and irrationals keep their visual character. Eg. *The √5 "is" the diagonal of a double square.*

The Grid makes it possible to construct, understand, demonstrate and remember any geometric figures quite easily.

2 – The vesica piscis

How does one build a grid with a rope, this ancestor of the compass and the rule ?

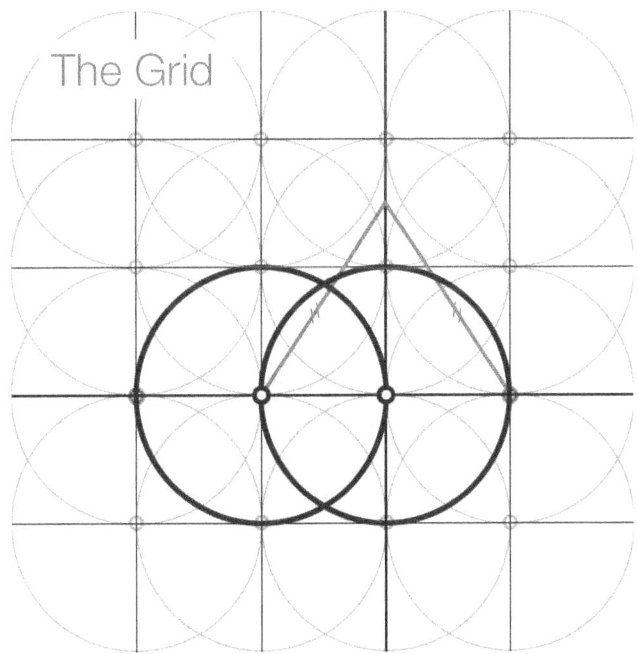

Fig.11 : The Vesica Piscis, "modulus" of the grid

The vesica piscis is formed of two similar circles, when the center of the one arises on its twin circle. It allows a mesh of the plan by a grid. This figure is probably the oldest.

Fig.12 : The "modern measures" of vesica piscis

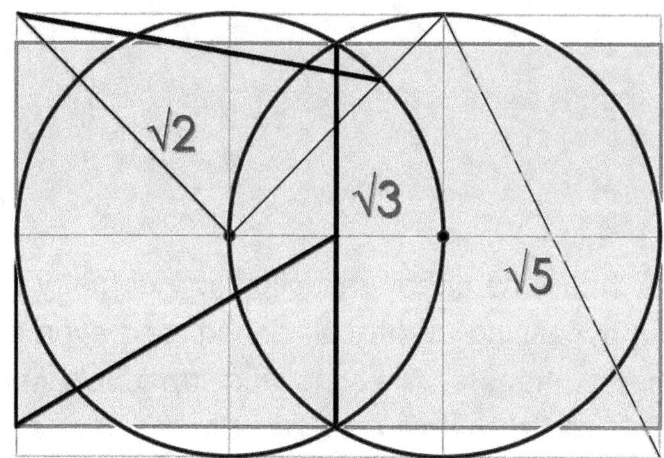

Considered as sacred by the Pythagoreans, this archaic motif is the original and sacred symbol of Venus. It appears in the Neolithic and seems to evolve between the notions of fertility / fertility and femininity. The explicit vulva of the almond is called "deic", and this is quite legitimate. The female Latin nouns *Vēsīca, Vesica, vesicæ* translates as bladder, but also refer to the female vulva.

3 – The Solar / Zodiacal Wheel

Draw a circle of diameter 2 in the center of the grid of 4 by 4. Its curve crosses the grid every 30° (one twelfth of full angle) and these points define a partition of the circle in 12 equal sectors

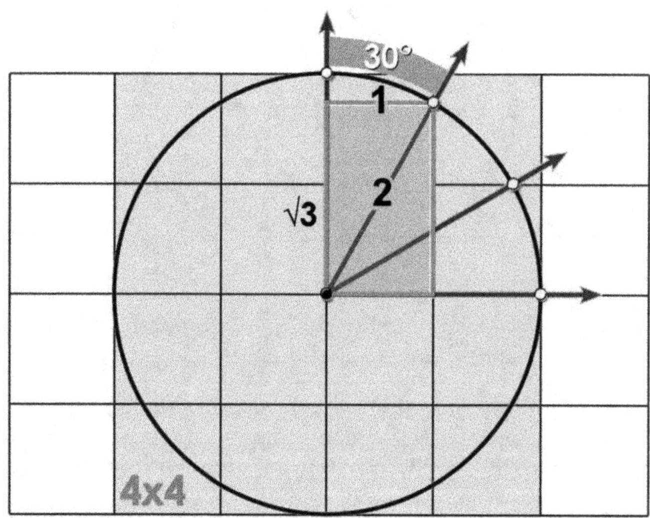

Fig. 13 : The circle of 4 on a square of 4x4

4 – The hexagram and family Δ

This logic is at the origin of the hour dial (day), and the wheel of the zodiac (year).

Subtle relationships appear, combining angles and measures. For example : A rectangle of **type Δ** (of proportion √3) contains three of the same type. We will come return to this property later.

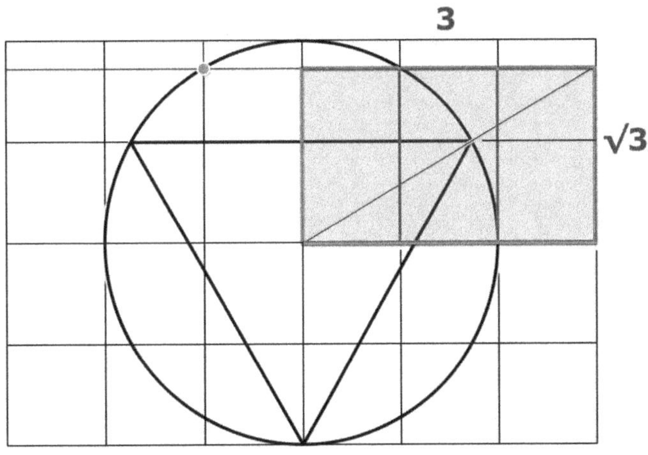

Fig.14 a : The rectangle of type Δ

The Seal of Solomon

This is one of the first figures with the rope. The center of gravity of the two triangles Δ (of water and fire), and the centers of their inscribed and circumscribed circles coincide.

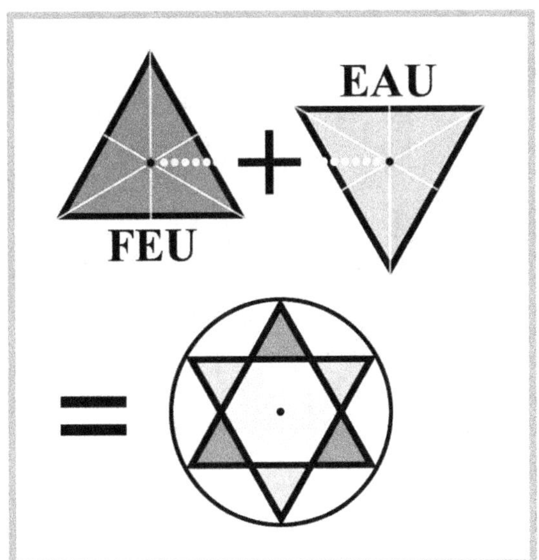

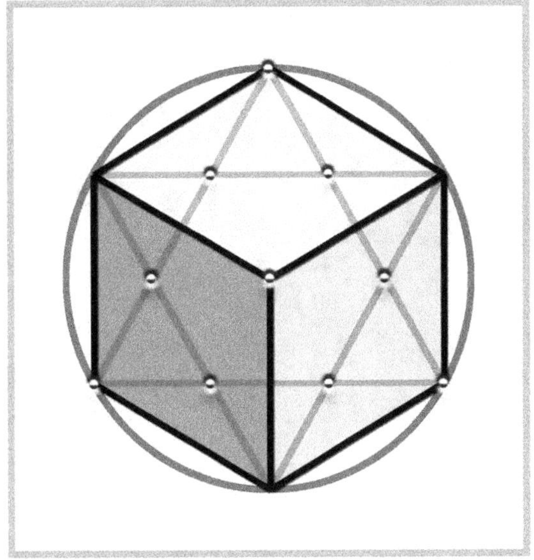

Fig. 14 b : Solomon's hexagram and its developments

The star is marked by the 6. Without presuming the sense given to it by the Egyptians or the Pythagoreans, the "ideas" which are attached to this number are, according to what is called tradition, although recent: Love, Association, fidelity, balance - Eternal / cyclical, Sun - surface of the sacred triangle, and of course hexagram.

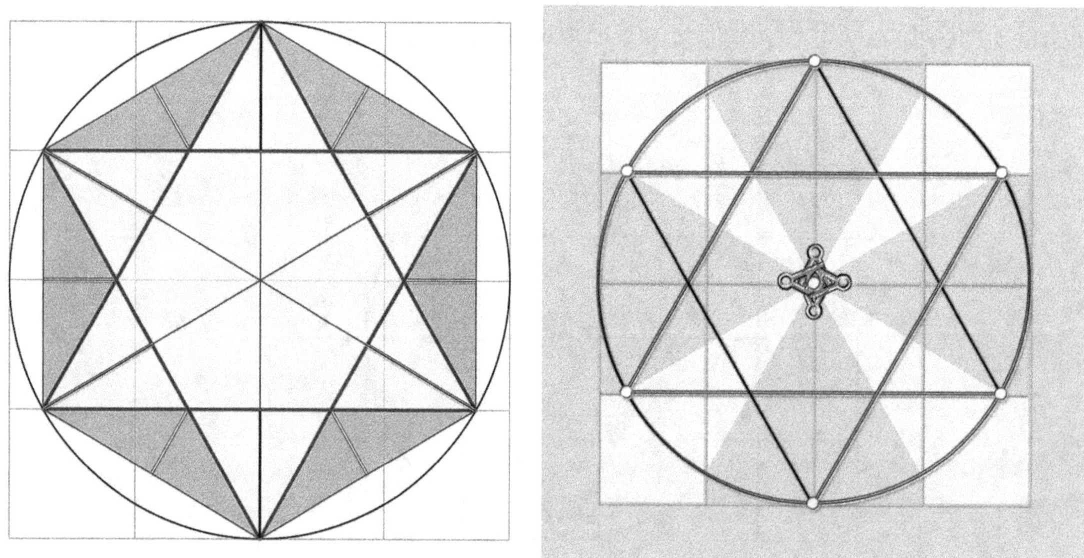

Fig. 14 c : Decompositions of the hexagram and the hexagon

The circled cross emphasizes the relations between the logic of the three and that of the four by the play of the circle.

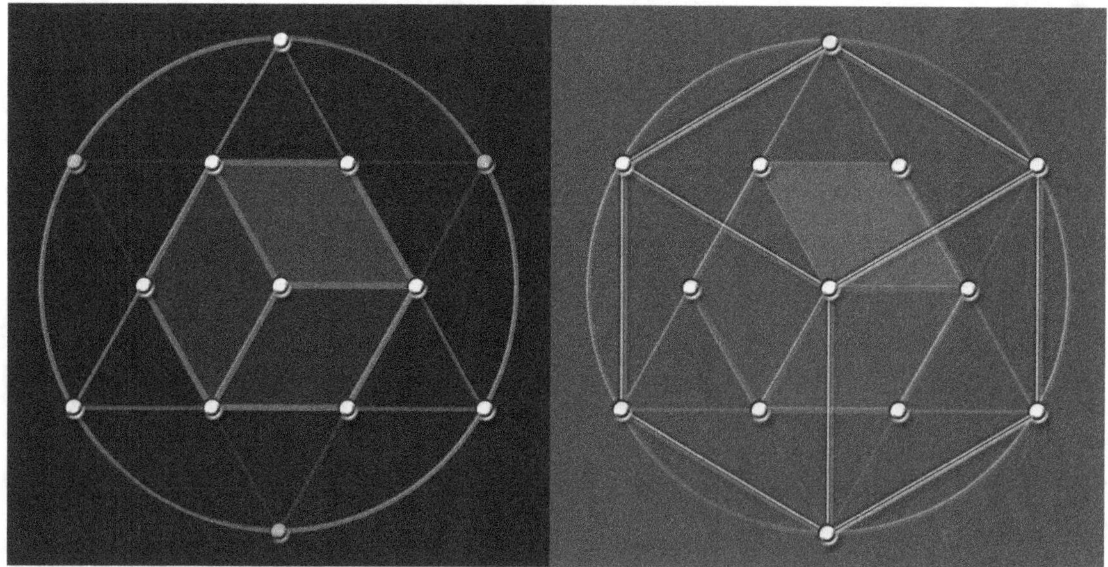

Fig. 14 d : The Tetraktys and the hexagram

The Pythagorean Tetraktys (or Decade) was revered as a structure that sums up the world. All dimensions of space are represented: 1 (point), 2 (line), 3 (surface) and 4 (solid).

The Pythagoreans have thus made of ten the most perfect of numbers. It carries within it all their differences. "Every reason, every proportion, every numerical form is contained in the decade." (Porphyry, Life of Pythagoras, § 51-52).

The cube is in the heart of the Tetraktys, which is built twice, from top to bottom and from bottom to top. In total the figure has 10 + 3 = 13 points, which reminds the triangle 3-4-5 and the rope with 13 knots.

The hexagram and vesica piscis

The vesica piscis is closely related to the equilateral triangle, the figures are part of a large family that can be called Δ. The links find different expressions.

Fig. 14 e : The "tuning fork" of the Byzantine grid

Here is the leading figure of the Byzantine grid, which Andrei Rublev used to build his "Holy Trinity". It plays the role of tuning fork for the whole of the work. The hexagon inside the hexagram takes the measure of the vesica, whose almond measures 2 in height. The circle of the vesica measures accordingly 2/√3 of radius. The almonds of the vesica at 45° are on the hexagram.

N.B. : As we might expect, the Byzantine grid is not a trivial grid, even governed by the golden ratio...

5 – The pentagram and the vesica piscis

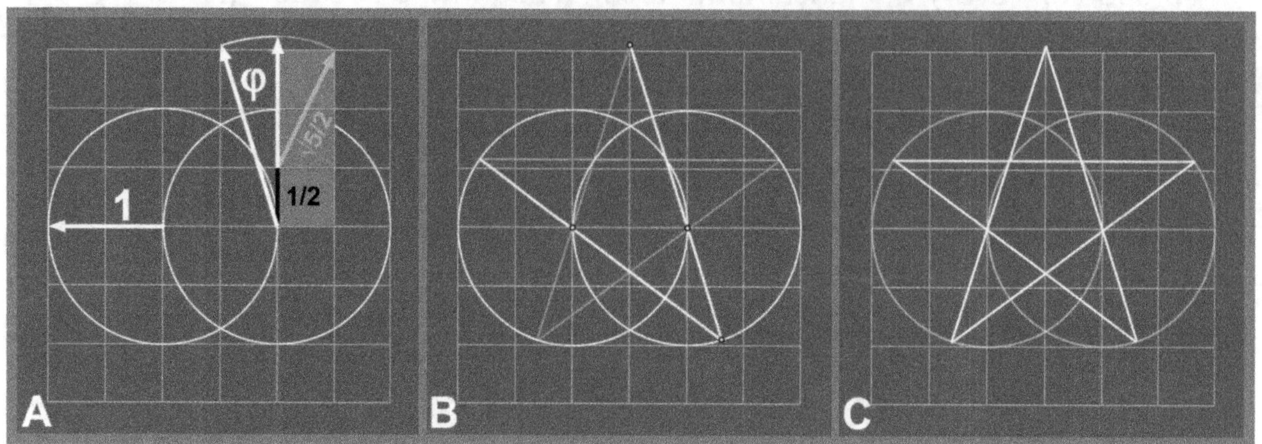

Fig. 15 : The grid is here doubled: a tile is therefore 1/2.

The pentagram is marked by the golden ratio. For a symbolic relationship to be revealed, a coincidence of figure must be found. Here it develops from the center of the circles of the vesica.

A – We find the point on the central axis, such as $\varphi = 1/2 + \sqrt{5}/2$

This sum results in a tile plus the diagonal of a double tile. This total is then transferred to the vertical which traverses the almond in height.

B – It is now sufficient to extend the bar of this golden ratio up to the circle to obtain the complete branch of a pentagram. The addition of 1 unit by the circle establishes its center as the point of intersection of the branches (this is the point of separation according to the golden ratio). Then, from the bottom point, you go through the symmetrical center, on the left, of the vesica.

C – These demonstrations use only the axioms of similar triangles. With this modest material, our distant ancestors were able to conceive a certain number of properties of the pentagram.

N.B. : 8 of the 10 points of the pentagram are on the circles.

6 – The Right-Angled Triangle

Fig. 16 a – *With the Eyes* – Any right triangle is a half rectangle. According to this, the middle of its hypotenuse is the center of its circumscribed circle, where diagonals intersect. The reverse is also easy. From a segment becoming hypotenuse, every point of the circle that takes the segment for diameter (and therefore its middle for center) is the point of a right angle with both ends of the segment. It is enough to reconstitute the rectangle to see it.

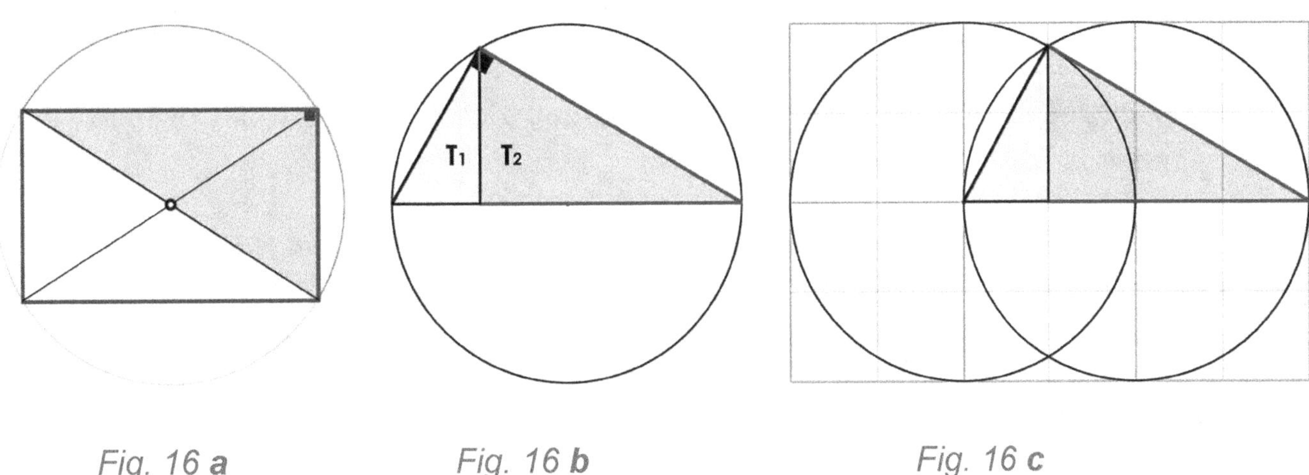

Fig. 16 *a*   Fig. 16 *b*   Fig. 16 *c*

Fig. 16 b – *With the Eyes* – The next step is to put the hypotenuse horizontally and draw the height of the right angle. The two triangles separated by the height, T1 and T2, are similar. Indeed, their three sides are found at right angles, starting with the hypotenuse (rotation of a quarter turn). This simple observation will allow us to connect the Vesica Piscis to the 3.

Fig. 16 c – *With Calculation –*
The grid is here doubled to facilitate the reading of the figure.
Let k be the measure of the height.
(Half the height of the almond).
> light grey triangle: width 1 and height k
> dark gray triangle: height k and length 3

The dark gray triangle has the same proportions as the light grey.
> —> k / 1 = 3 / k   So k.k = 3

Other readings of the same figure are possible,
Perhaps more in the spirit of original geometry.

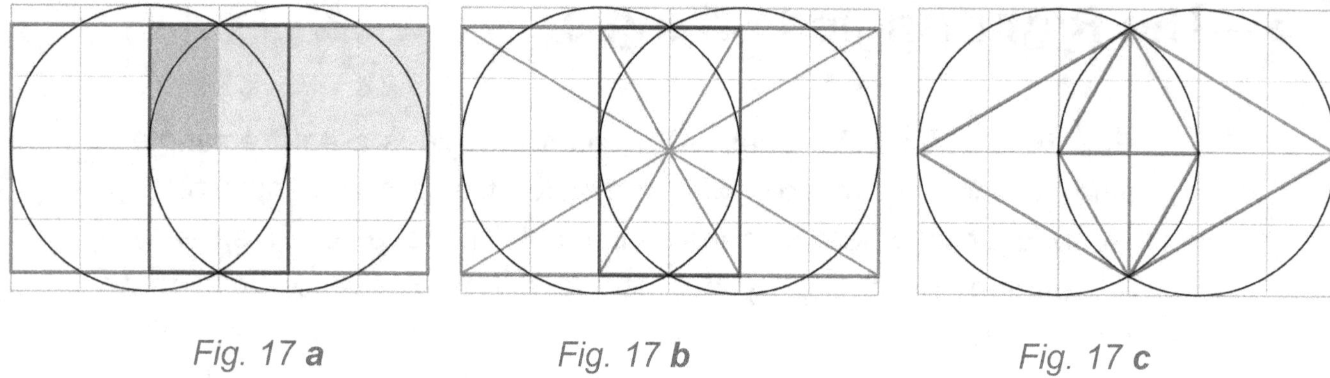

Fig. 17 *a*   Fig. 17 *b*   Fig. 17 *c*

Based on the previous visual, one can highlight similar rectangles, each representing a quarter of their reference rectangles : that of the almond, and that of the height of the almond over the total width of the two circles united. **NB:** The ratio of this large width to the width of the almond is 3.

In the two other cases, the diagonals of the two rectangles constitute two equilateral triangles. Perfect expression of 3.

Fig. 17 c : These equilateral triangles can be placed in a different way to form a lozenge. We have also seen that the angle of the hypotenuse of the triangle T1 (inside the almond) is also related to the 3 : it is the partition into three equal parts of the right angle.

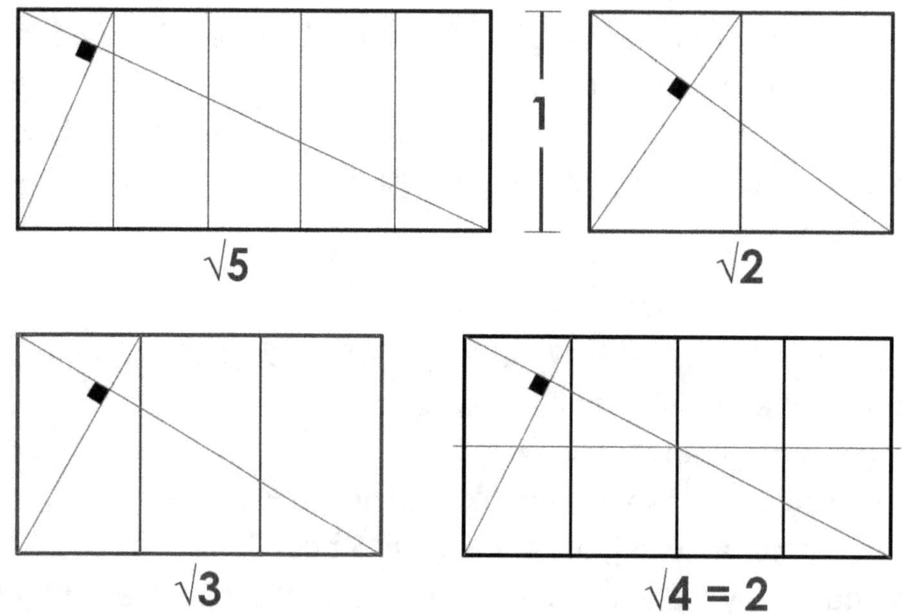

Fig. 18 : The "roots of integers" as proportions

In any rectangle of proportion √N with N integer,
there enter exactly N rectangles of the same proportion

III – THE TRIANGLE 3–4–5

This chapter shows the extraordinary coherence of geometry with the eyes. In current scientific terms, logical arguments are limited to the axioms of similar triangles - to which Thales will give his name. In common language, visual evidence is enough to explain everything (in order to understand), and this for all the figures. And for good reason : these mathematics developed before calculation (and writing). Raphael Legoy proposes to qualify these approaches as **monstrations**.

A number of prejudices are seriously compromised by this body of evidence. Certainly, the Neolithic men had a practical mind. However, they deserve equal consideration on the theoretical level. The geometrical knowledge which we discover has not an empirical and approximate character of which it is commonly considered. In general, as soon as they awoke to mathematics, men adopted a state of mind much closer to ours than was suspected : that of evidence, of remarkable properties, and their overall view. This wide culture arouses admiration, interest, and invites to meditation.

1 – The sacred triangle

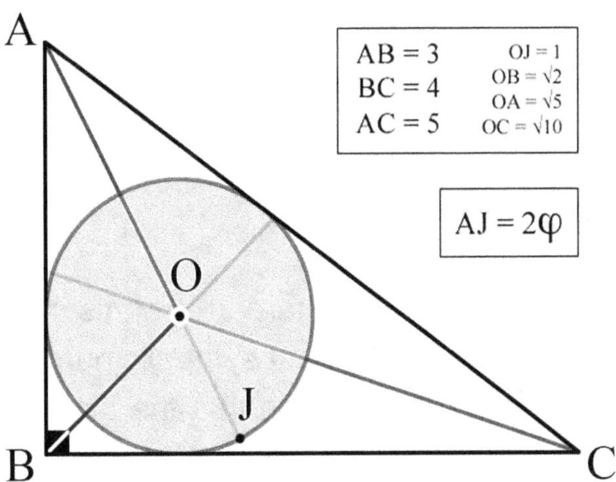

Fig. 20 : The triangle 3-4-5

The triangle 3-4-5, also called sacred triangle, Egyptian triangle, triangle of Isis, or triangle of the surveyor, is the queen figure of sacred geometry. It carries in it all the numerical values of the symbolic: the integers from 1 to 7 plus the golden ratio (φ) and the root of three ($\sqrt{3}$). It gives its full meaning to the grid, and to its unit.

The principal monstrations which follow are unpublished. Some relate to properties that history has forgotten (or ignored) for centuries. This is the case for the golden proportion of the triangle, which has remained hidden on one of its bisectors. (Note: When the author is not specified, the monstrations are signed Yvo Jacquier.)

2 – The first three measurements

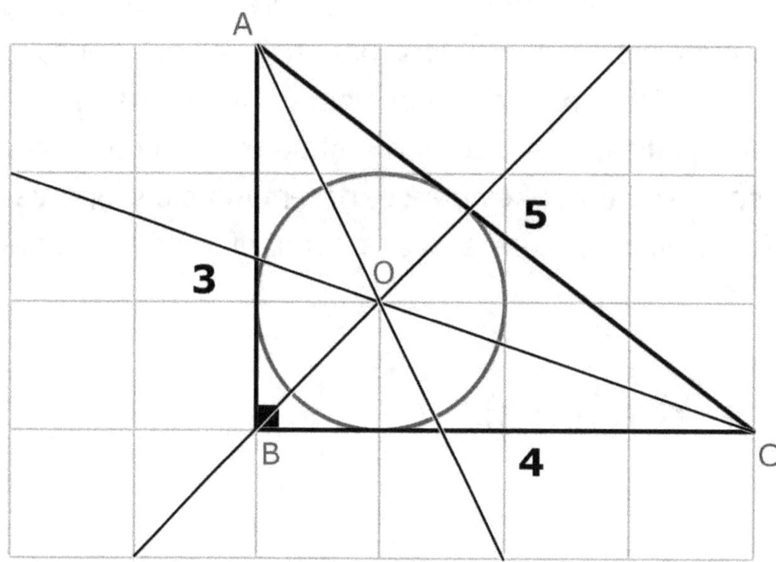

Fig. 21 : The measurements of the sacred triangle

The sides of a right-angled triangle that form the right angle are called *cathetus*, facing the hypotenuse.

The three bisectors are the natural diagonals of a single, a double, and a triple square. The sum of this first symbolic number, called "order of the bisector", and the length of the side that the bisector passes through is always 6, namely : 1 + 5, 2 + 4, et 3 + 3.

The length of the third side, AC, seems axiomatic. It took years to resolve a point that proves to be crucial.

Depending on whether the length of the 5 is admitted or demonstrated, the pre-Euclidean geometry does not enjoy the same status, particularly in relation to that which the Greeks will construct. Can this thus be considered as a scientific or as a some « do-it-yourself » process ?

First monstration

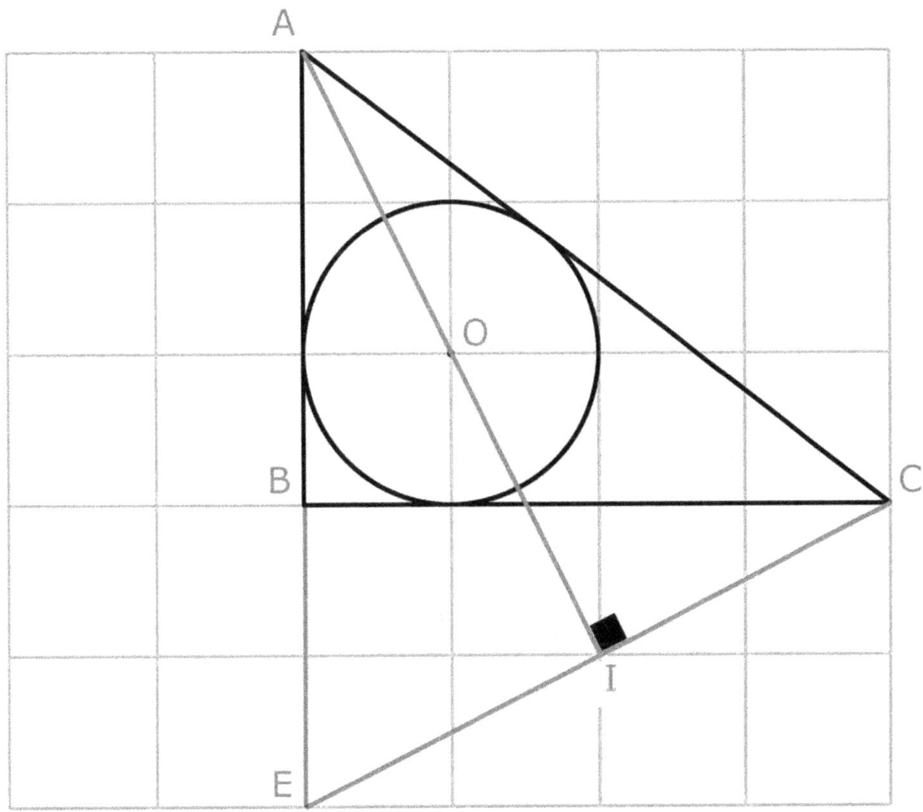

Fig. 22 : The symmetrical triangles in the monstration of the "5"

Let a triangle ABC, with AB= 3 vertical, and BC= 4 horizontal.
Let the point E, located at two tiles under the point B : AE = 5.

According to the grid, EI = IC = ∂
(with ∂ diagonal of a double square)

Now AI is orthogonal to EC
IC is a rotation at 90° of IO, according to fig. 10 b)
Therefore AIE and AIC are symmetrical.
Therefore AE = AC = 5   *What Was to be Shown*

N.B. : The inscribed circle of the triangle does not intervene in this monstration. Yet it reveals its measurements. Then, the first six integers are symbolically united by one same figure, including the 6 of the surface of the triangle.

Other monstrations establish the 5 of the hypotenuse, but they use a property of the inscribed circle of the triangle: its radius is equal to 1.

3 – The inscribed circle, *called « intimus »*

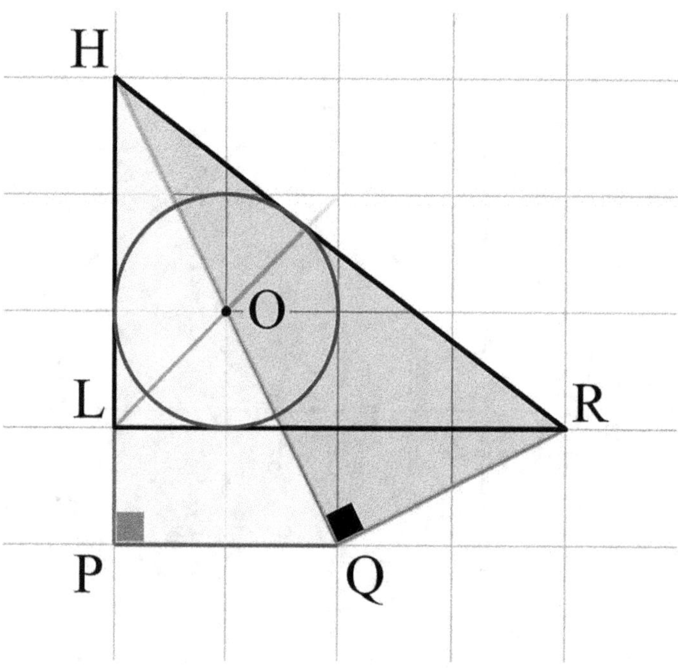

Fig. 23 : The monstration of the inscribed circle of the triangle 3-4-5

Let HLR, a triangle 3-4-5.
Let the points P, Q, et O on the grid,
easily identified in the figure.
The triangles HPQ and HQR are similar.
According to what the angle PHQ is equal to the angle QHR.
Therefore O is on the bisector of the angle in H of the triangle 3-4-5
Now, O is also on the bisector of the angle in L.
The center of the inscribed circle of the triangle is therefore at the intersection of the three bisectors.
Two are enough, so it is O,
And the radius of the circle is therefore 1.   *What Was to be Shown*

4 – Pinwheel tilings

The coefficient that differentiates the triangles is HR/HQ = HQ/HP = QR/PQ = √5/2 ≈ 1,118. I thought at first that Egyptians did not care about this kind of thing.

Jean-Paul Mercier shows that if they did not identify this irrational value (√5/2), the Egyptians were perfectly capable of practicing its square value : 5/4 !

It is enough to cut each of the triangles of the previous figure according to a "template" known as the triangle of Pinwheel (Charles Radin thus defined a method of paving aperiodic, from a construction of John Conway). For the Egyptians, it is half a double-square. According to this subtle cutting, which plugs its holes with their symmetrical surfaces, the triangle HPQ contains 4 units and HQR : 5 ...

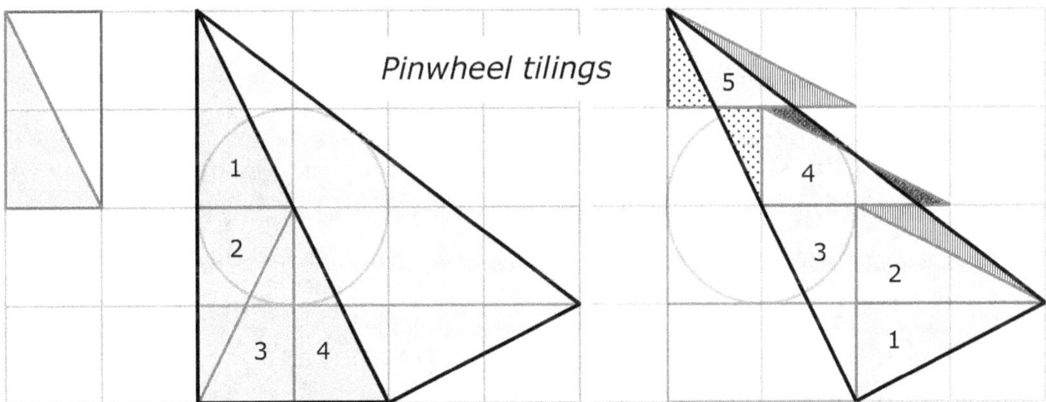

Fig. 24 : The cutting of the surface according to the triangles of Pinwheel

5 – Others monstrations of the 5

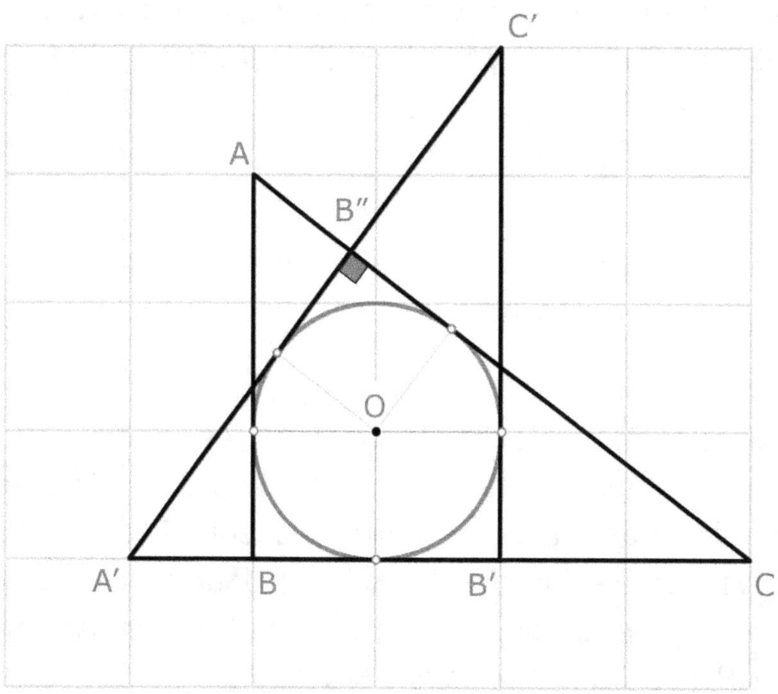

*Fig. 25 : Monstration by **Frédéric De Ligt***

Monstration 2 of the hypotenuse

Let the triangle ABC,
with AB vertical = 3 and BC horizontal = 4
Let O be the center of the inscribed circle of the triangle
Let the triangle A'B'C', rotation of ABC around O at 90°
in the counterclockwise direction.
Let B" the point where the hypotenuses of the two triangles intersect.
(90° is the right angle, between the hypotenuses)
The angle in B" of A'B"C is therefore a right angle.

A'B"C borrows an angle to ABC and another to A'B'C'
The triangles ABC, A'B'C' et A'B"C' are therefore similar.
The inscribed circle of A'B"C' stays the same.
The three triangles are isometric (same measurements).
Now, as we have shown, the inscribed circle has the diameter 2.
Therefore :
The hypotenuse of A'B"C', like that of the other two,
measures A'C = 1+ 2 + 2 = 5.

What Was to be Shown

Monstration 3 of the hypotenuse

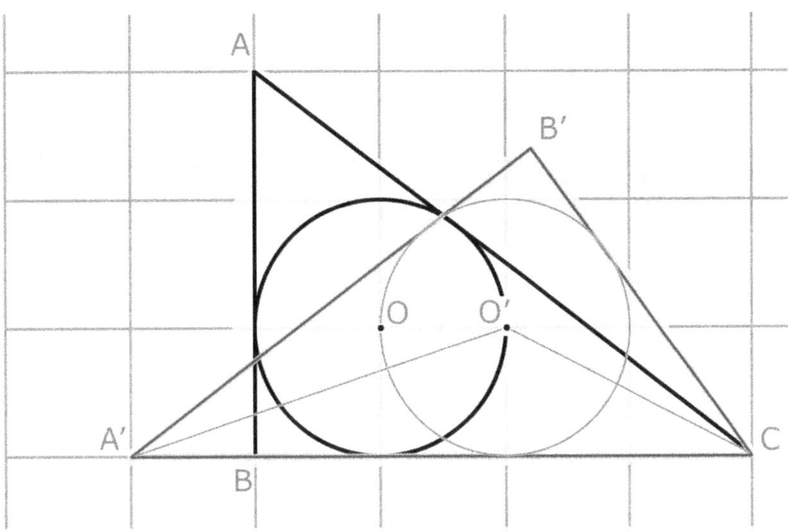

Fig. 26 : Monstration with vesica piscis

This monstration allows us to understand the fascination of the
Ancients for the figure of the vesica piscis.

Let the triangle ABC, with
AB vertical = 3 et BC horizontal = 4.
Let O the center of the inscribed circle.
Its radius is 1 (established property).
Let the triangle A'B'C, built on the base A'C = 5,
taking the twin circle of the vesica piscis as inscribed circle.

The left side A'B' is constructed by doubling the angle of the diagonal
of the triple square which passes to the center O'.
Similarly, the right side CB' is constructed by doubling the angle of
the diagonal of the double square which passes through O'.
The two triangles have the same two acute angles
=> Therefore they are similar
Then the size of their circle is the same
=> They are therefore equal.
The small sides of the triangle A'B'C are 3 and 4 and especially :
The long side of the black triangle is 5.

What Was to be Shown

IV – THE GOLDEN RATIO

1 – The different approaches to the golden ratio

The origin and the appellations of the golden ratio

The mathematical symbol φ of the golden ratio honors the sculptor Phidias, representative of the first classicism in Ancient Greece. In 460 BCE, his sponsor and friend Pericles entrusted him the work of the Acropolis (including the Parthenon). His mastery of proportions is so exceptional that he takes into default the inhabitants of Athens. Laid on the ground, the representation of Athena seems very awkward, but once perched on its pedestal, the work of Phidias became divine ! At the end of his life, the Godfather of the golden ratio is victim of an ill-trial, and his jealous contemporaries force him into exile - to the city of Olympia.

Much later in the twentieth century, in the years 10, the critic and British fencer Theodore Andrea Cook (1867-1928) decides with a friend, a US mathematician named Mark Barr to propose the notation φ (the Greek letter Phi) as the mathematical symbol of the golden ratio, as a tribute to Phidias. The double argument of the consonance of the letter φ with that of π , much as with the name of the sculptor - made famous for his mastery of the golden ratio, is reported by Cook in his book « The Curves of Life »*. He makes the account of spiral formations in nature, science and art. And he particularly refers to the work of Leonardo da Vinci. Later, the term « golden ratio » will be set by the Romanian prince Matyla Ghyka in the 30s. (*) *Cook, Theodore Andrea, The Curves of Life (1914), Courier Dover Publications*

Here is an overview of the terms used about it :
1 - Number outrageous since irrational   Plato
2 - Extreme and mean ratio   according to Euclid
3 - Proportion of Euclid   Fibonacci

4 - Golden Section   (*sectio aurea*)   according to Vinci
5 - Divine Proportion   according to Pacioli
6 - Golden section   (*der goldene Schnitt*)   Zeising
7 - Golden number   set by Ghyka
8 - Phi   (φ - mathematical expression)   Theodore Cook
9 - Golden proportion   common usage

In english :   Golden ratio   Golden proportion
In german :   Der goldene Schnitt   Golden section
In czech :   Zlaty rez   Golden section

More : http://www.contemporary-painting.com/golden-ratio-history.html

The contemporary approach to golden ratio

The golden ratio is presented today as the positive solution of the equation : $φ2 = φ + 1$ with $φ = (1+\sqrt{5})/2 ≈ 1{,}618\ 033\ 988\ 7...$

φ is an irrational number (it does not correspond to any fraction of integers). On the other hand, it is not transcendental (as π). In concrete terms, it is accessible to the ruler and the compass.

This number is the subject of much speculation. In the field of mathematics, its most recent developments go through calculation, as evidenced by this equation borrowed from Wikipedia   [Fig. 27]

$$\varphi = \sqrt{1+\varphi} = \sqrt{1+\sqrt{1+\varphi}}\ \ \text{et}\ \ \varphi = \sqrt{1+\sqrt{1+\sqrt{1+\sqrt{1+\cdots}}}}.$$

These developments confirm the power of the golden ratio. On the other hand, they do not reflect its purely geometrical origin, as we shall discover. Secondly, the main disadvantage in terms of consequences of these developments lies in the lesson that is drawn from them clumsily : the golden ratio is conceived above all as a proportion, a sort of multiplication rule that produces beauty by reproducing/repeating itself. It was not the project of the mathematicians, for example of Fibonacci ... *And it is not enough to enclose an object (even an idea) in a golden rectangle so that it turns into gold.* The "canons" principle, so valued by history of art, is not in the minds of the ancients, those men of knowledge (especially artists and architects). These errors do not account for the level of geometry developed by artists and architects throughout the millennia (the Renaissance will be the apotheosis of this great culture).

Kepler's Third Law

This sort of mutation of φ, from its initial neolithic status, just according to the eyes, to the "modern" conception of the number, has evolved in stages. Kepler himself used φ to progress in his research !

$$\textbf{VENUS}$$

$$T = \frac{1}{\varphi} = \varphi\text{-}1$$

$$T^2 = D^3 = \frac{1}{\varphi^2}$$

Fig. 28 : *Golden relations between period and distance of Venus*

As Christophe de Cène explains in his article Kepler, from ancient knowledge to modern physics, Kepler revealed his third law in the face of an equation which associates the period T and the distance D of the planet Venus with Number φ.

This law, chronologically the first, marks the birth of science. The equation "$T^2 = D^3$" will be generalized to all the planets of the solar system.

The golden ratio by the ancient Greeks

The Greeks are primarily geometers. It would therefore be unfair to assign the misconception that we have today of the Golden Ratio. For proof, the proposition 11 of Book II of Euclid's Elements, that Jean-Paul Guichard reminds us.

For Euclid, the purpose is to share a segment into two segments so that the ratio of largest to smallest is equal to the entire segment to the largest. Note : This proportion (in extreme and mean raison) is what we call the golden ratio.

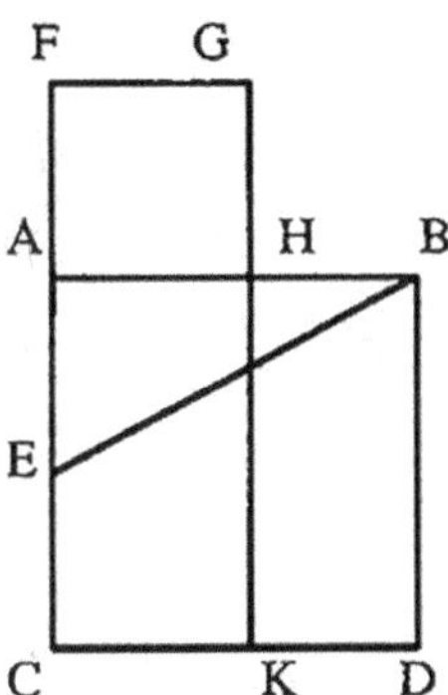

Fig. 29 : The figure of the Elements

- The segment to be shared AB (AB = 1)
- H the point as AH / HB = AB / AH (= φ)

with the construction of the point H :

- ABCD square built on AB
- E middle of AC (AE = EC = ½)
- EF = EB (= √5/2, and CF = ½ + √5/2 = φ)
- AH = AF (= φ - 1)
- AFGH is a square

Then Euclid shows that :

- Area of FGHA = area of HBDK

$[(φ - 1)^2 = (2 - φ)×1$ where $φ^2 = φ + 1]$

where $AH^2 = HB×AB$ or AH/HB = AB/AH (= φ).

Subsidiary Note : we can complete the golden rectangle FRDC

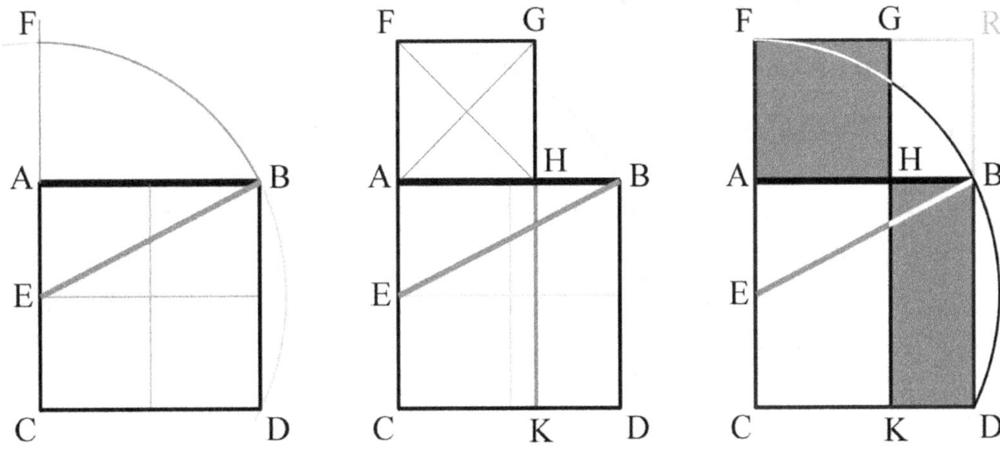

Fig. 30 : Decomposition of Euclid's demonstration

In the resolution of this problem, Euclid anticipates the following. In fact, there is identity of the small square FGHA and the residue HBKD, resulting from the removal of a golden rectangle AHKC to the large square ABDC. Didactic !

The golden ratio and proportion Δ (√3) for Egyptians

Egyptian geometry is the perfect example of the accomplishment of geometry with the eyes, built on a grid without calculations, without equations and without writing. However, the Egyptians are not alone in this practice. We must cite the Mesopotamian current (Sumerian and then Babylonian); It probably shares the same origin, megalithic. And others more distant, in India and China.

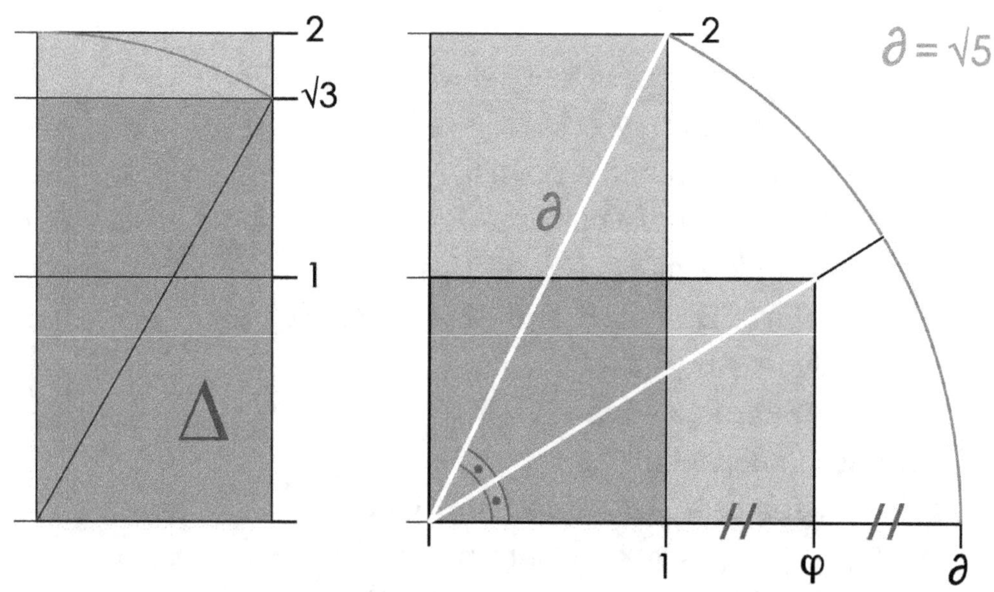

Fig. 31 : The proportion of type Δ (√3) and of the golden ratio (φ)

Left : The root of three is obtained by folding the 2 of the double-square which becomes diagonal of a rectangle qualified of type Δ.

We build indeed a right-angled triangle of measurement 1 and √3, with 2 for hypotenuse. And this is half an equilateral triangle (hence Δ).

Right : The golden ratio is obtained by bisecting the angle of the diagonal of a double square. The bisector crosses the horizontal of the first square at the distance φ. It is also the average of 1 and ∂, diagonal of the double-square (= √5).

These two numbers are born from the same double-square by a set of diagonals. This common origin is the sign of a relationship that the √3 confirms, visually, thanks to the square.

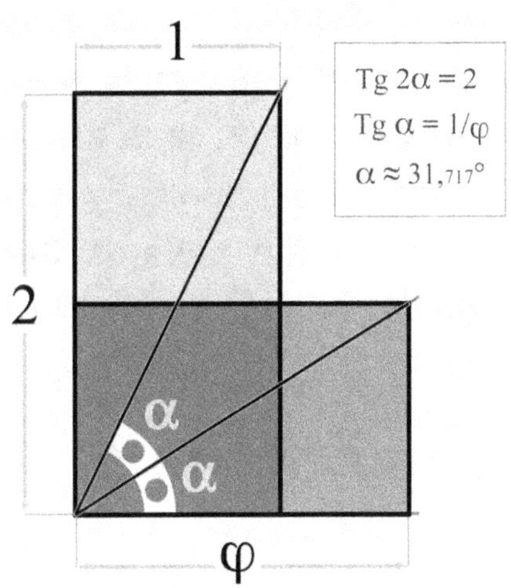

Fig. 32 : The golden ratio with the eyes

This visual shows the simplest definition of the golden number. *The wide angle of the diagonal of a double square measures twice the small angle of the diagonal of a golden rectangle.*

The first attitude towards this peculiarity of the number of gold was to make it a property. Conversely, the most well-known geometrical properties of the gold number were chosen to define it : *In a golden rectangle, the removal of an inscribed square produces a residue of the same proportion.*

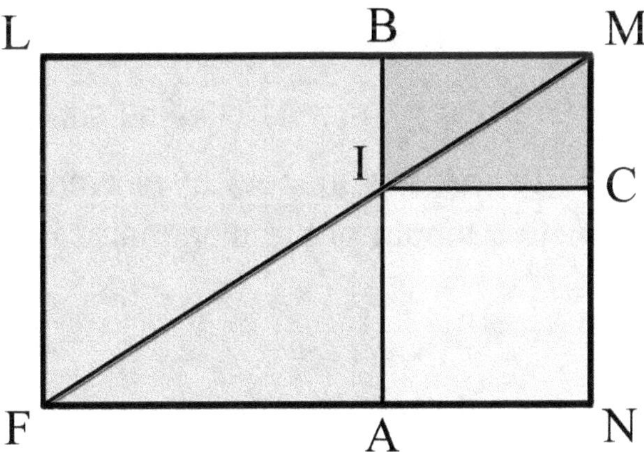

Fig. 33 : Other property of the Golden Proportion

This alternative definition is also very simple and can be understood as follows : The rectangle LMNF is a golden rectangle if the withdrawal of the square LBAF produces a second golden rectangle BMNA. This one, by removing a square ICNA produces a third golden rectangle IBMC. The diagonal of the large golden rectangle is also the diagonal of the smallest, and the point I is located on the diagonal.

One can go from one definition to another, then the second is a property, a consequence of the first. However, several considerations incline to think that the first definition was that of angles. On one hand their practice and their mastering logically preceded the exercise of proportions, which are a first step towards the calculation. Then, the definition of proportions does not allow direct construction of φ.

2 – "Egyptian" monstrations of the Golden Ratio

The pinwheel triangle

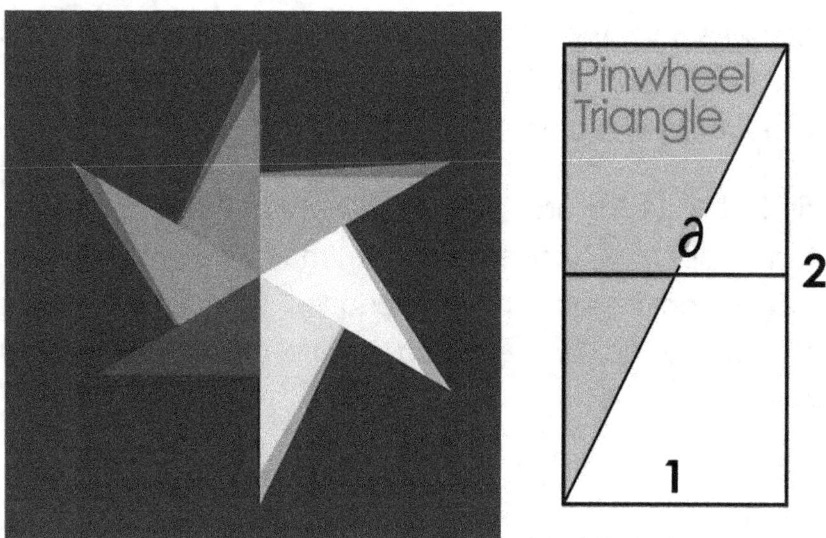

Fig. 34 : The pinwheel triangle, half a double square

A figure will be useful: the triangle of pinwheel. This is half of a double square. We note ∂ length of the diagonal, since the √5 is neither useful nor revealed..

The grid as a scale

We choose intentionally a grid of two tiles for the initial square.

First, to allow the easy use of the triangle of pinwheel in the monstrations. Then we will meet a golden rectangle of this size in the triangle 3-4-5

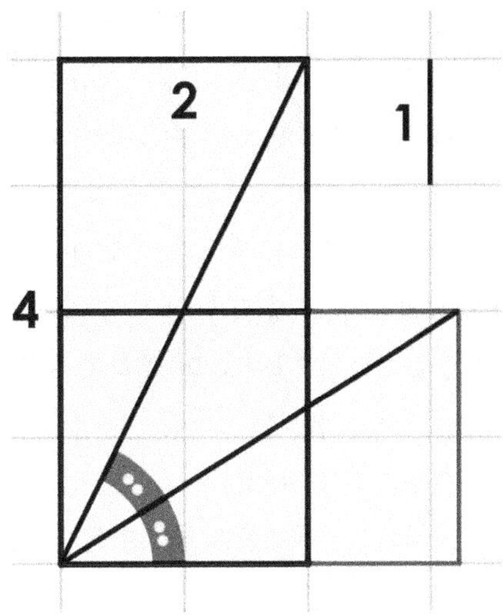

Fig. 35 : *The golden rectangle on its ideal grid*

It is with this scale that we will study the golden rectangle. Let us take again this figure, adopted as a definition. We shall deduce from it the measurements of its golden rectangle and the property of the proportions which we have stated.

The first kite - Measurement of the golden rectangle

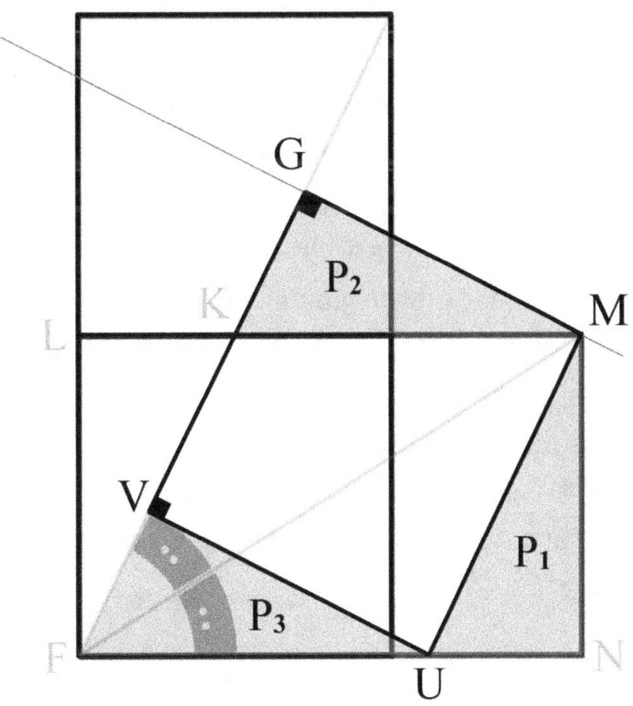

Fig. 36 : *Appearance of pinwheel triangles*

— Build P_1, a first pinwheel, relying on MN to find the point U.

— Trace the perpendicular to the diagonal of this double square, through M, and then find the point G on this line, as GF orthogonal to GM.

— Same process at the point U : trace the perpendicular to the hypotenuse of P_1, which finds the point V on the line VF.

— GMVF is a rectangle by construction.

— Call KGM : P_2 and FVU : P_3.

LMN and GMU are right angled in M,

Therefore the angle in M of P_1 and P_2 is the same.

More the both have a right angle,

therefore their third angle is the same.

P_1 and P_2 are similar

Their three sides are parallel, so P_2 and P_3 are similar.

More, in this particular case, P_2 is the translation of P_3 ,

so P_2 and P_3 are isometric.

F, V, K et G seem to be on the diagonal of the great double-square.

Indeed, by construction, they are on the same line.

Then P_3 is similar to P_1 and P_2, with P_1 triangle of Pinwheel.

So the medium angle in F of P_3 is the great one

of the diagonal of a double-square. Therefore :

F, V, K et G are on the diagonal of the great double-square.

Then, according to the Egyptian definition of golden ratio,

FM is also the bisector of the angle in F of GFN.

The triangles FGM and FNM have a right angle.

The third is therefore equal, and they are similar.

The hypotenuse of these triangles is the same : FM.

FGM and FNM are therefore perfectly symmetrical.

Therefore GM = VU = MN = 2

(P_1, P_2 et P_3 are three pinwheel triangles)

FV = KG = UN = 1, and especially KM = FU = UM = ∂

The large measurement of the golden rectangle is therefore

FN = FU + UN = ∂ +1.

What Was to be Shown

Remark : FLK is also a pinwheel triangle.

Second kite - A known property

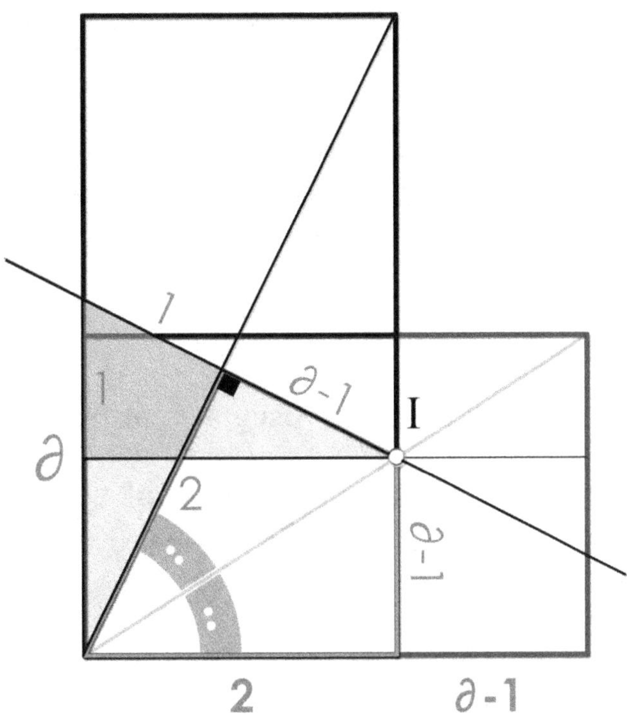

Fig. 37 : How to do without calculation by relying on geometry

This demonstration must establish a property that is often used as a geometric definition of the golden ratio. However, this proposal does not build the golden ratio, contrary to the definition by the angles:

In a golden rectangle, the withdrawal of an inscribed square gives a residue with the same proportion.

Cross two triangles of Pinwheel as shown in this visual. The first begins from bottom left and its hypotenuse is on the edge. The second extends the side one of the first, and retrieves the horizontal at the point I. These triangles will allow us to measure everything.

A second potential kite appears here, with its two triangles on either side of the FI line. Just see the measurements of the triangles of Pinwheel to find those that kite, whose triangles are symmetrical on both sides of FI.

The great measure of the large golden rectangle is $\partial+1$.

Simple calculation : $\partial + 1 - 2 = \partial - 1$

So, the point I is the corner of a square which side is ∂ -1 (right).

The angle of the kite in F is the complementary part of the tip of the triangle of Pinwheel to get a right angle. So, this is the wide angle of the diagonal of a double square. I is on the line bisecting this angle at F, so it is, according to the definition "with eyes", on the diagonal of the large golden rectangle.

The withdrawal of a square (side 2) to the large golden rectangle, leaves a first residual vertical rectangle (right) width ∂-1. The same operation, removal of a square this time aside ∂-1, produces a rectangle with the same diagonal than the great golden rectangle. This large rectangle and the smaller are similar, i.e. golden.

Now, how to show that the first residual is also a golden rectangle ?

Jean-Paul Guichard gives us the simplest solution. It suffices to rotate this rectangle : its inscribed square rotates of a quarter turn...

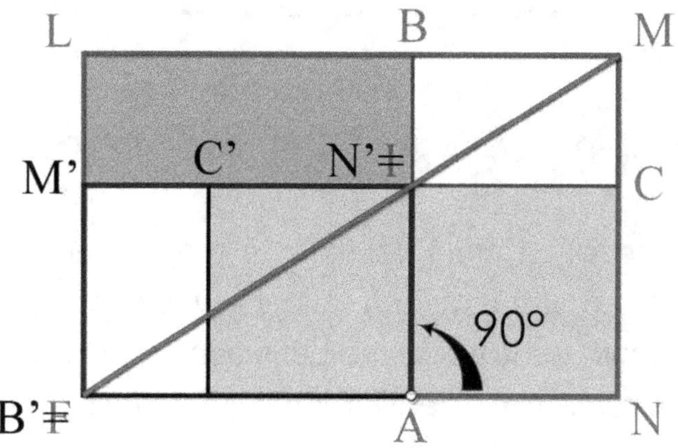

Fig. 38 : A rotation of the rectangle around point A

Reminder : LBAF is a square of side 2.

The rectangle ABMN rotates 90° around A.
Its dimensions are at the starting, height : 2 and width : ∂ -1.

This new rectangle is AN'M'B' with N' = N and B' = F

B'N' is none other than FI, the golden diagonal.
The rectangle AN'M'B' is therefore golden.

(argument of Chapter III-6).

What Was to be Shown

Correlates

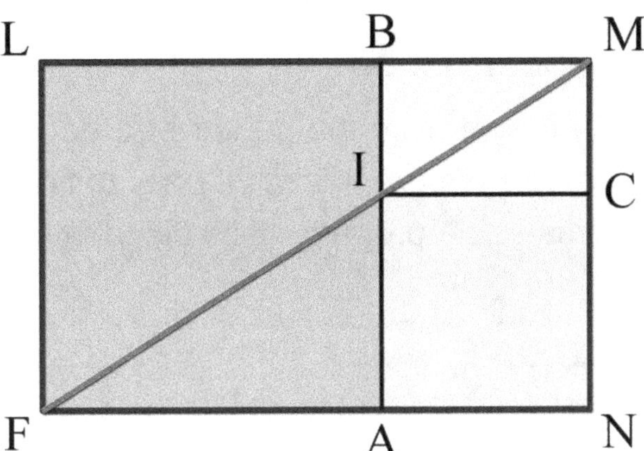

Fig. 39 : The most known property of the golden ratio

We find back the classical presentation of the golden rectangle, in which the removal of an inscribed square produces a residue with the same proportion. The rectangle ABMN, obtained by removing the square ABLF is similar to LMFM.

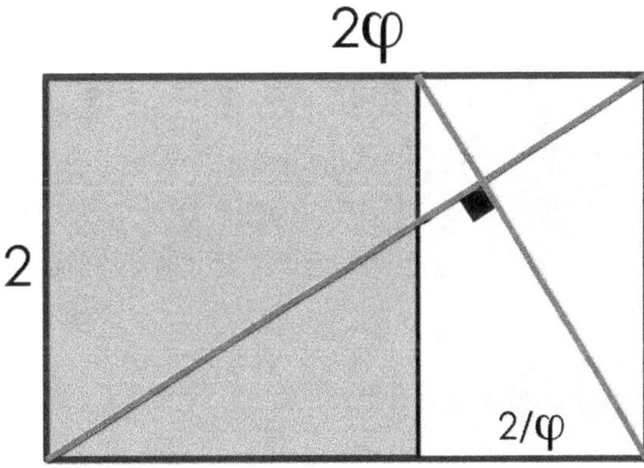

Fig. 40 : Orthogonality of diagonals of the rectangles

Correlate of this property : *The diagonals of a golden rectangle and its residual rectangle form a right angle.*

The two rectangles have the same proportions, so they share the same type of diagonal. Their orientation make an angle of 90°, therefore the diagonals also.

3 – Other remarkable property of golden rectangle

Simplification of the construction

A simple circle drawn from the center of a double square, through the four corners of double square, makes it easy to find the measure of φ, and it highlights a remarkable property of its diagonals.

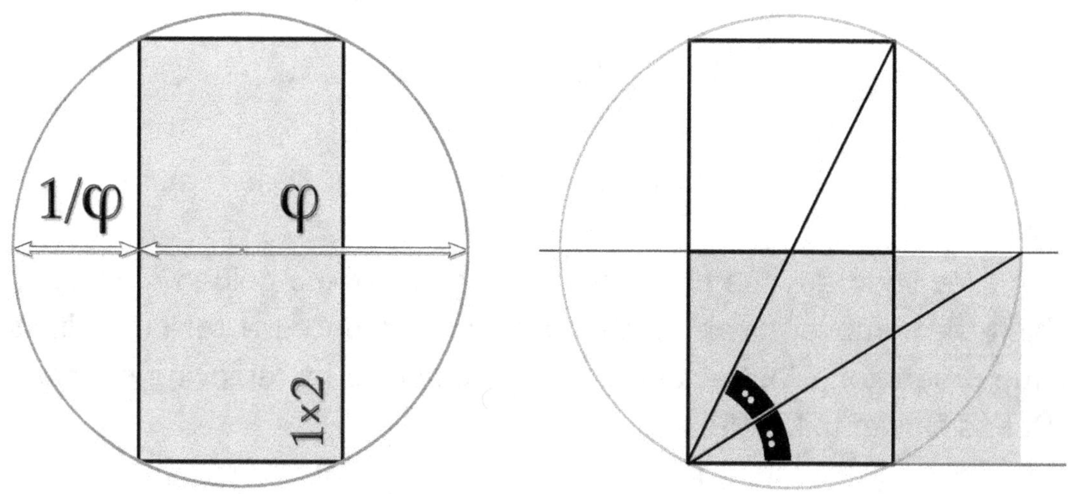

Fig. 41 : Simplest construction of golden ratio

The first golden rectangle that we built was $\partial+1$, measured on a dual grid. The golden rectangle of height 1 measures accordingly $(\partial+1)/2$, which is half of the hypotenuse of a triangle of Pinwheel plus half a unity tile. This is the case of the measurement shown on the left figure.

The construction of Albrecht Dürer

This construction shows a peculiarity of the diagonals of a golden rectangle (we will demonstrate this property in the following) :
*When a golden diagonal is horizontal,
the other is that of a double square.*

This means that : *The small angle (acute) between the diagonals of a golden rectangle, at its center, is equal to the wide angle of the diagonal of a double square.*

Albrecht Dürer uses this principle in the construction of his engraving « MELENCOLIA § I », 1514.

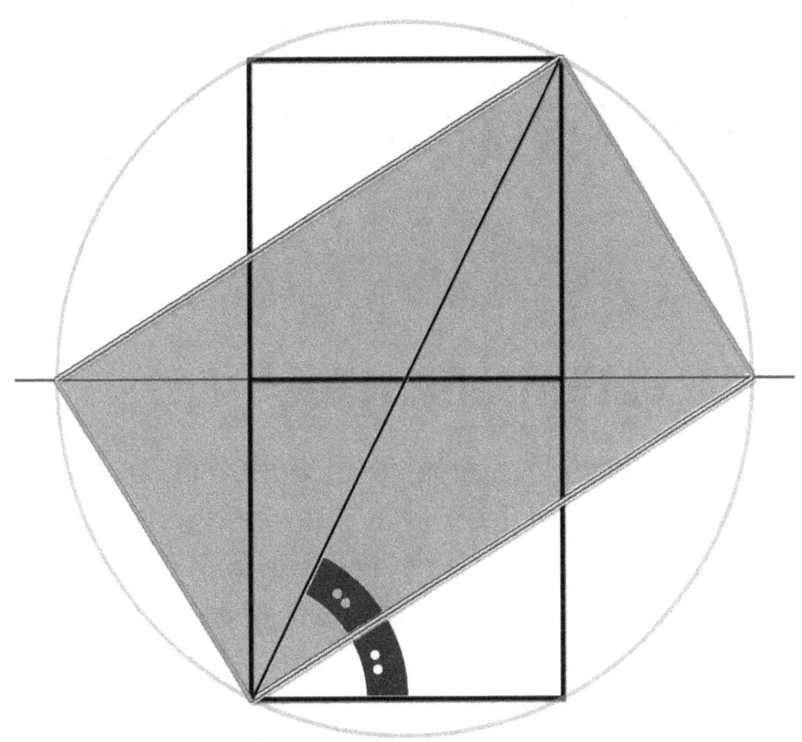

Fig. 42 : The figure of Albrecht Dürer

Albrecht Dürer (1472-1528) deserves a real tribute. Without his pedagogy, it would not have been possible for me to reconstitute this geometry. The artist of Nürnberg conceived what I called the "Didactic Project of Dürer", a device by which he explains step by step the functioning of the language of the image. From the associations of the subjects in the way of a puzzle to the most sophisticated geometric structures. This device includes a deck of tarot cards, known as "Tarots de Marseille" in its best version (Nicolas Conver), and four engravings of an unusual format gathered under the generic term of Meisterstiche. The most famous work of his time, Renaissance, is one of them : the mythical "MELENCOLIA § I", engraved in 1514. The 22 major arcana of the tarot are composed on the same grid as Melencolia. This physical compatibility makes it possible to place the cards on the engraving, so that their symbols are perfectly combined. The artists of the Renaissance needed an engraver to guarantee the durability of their will. (See XI, 6)

V – GOLDEN EXPRESSIONS OF THE TRIANGLE 3–4–5

1 – The consecration of angles

The geometry with the eyes is thought with angles rather than calculation, with similar triangles rather than the theorem of Pythagoras. The corpus we reconstitute ignores the concept of area as much as writing, and "calculation" is limited to counting distances.

For a long time, I believed that the people who had conceived this geometry - especially the Egyptians and the Mesopotamians, appreciated the angles with the eye of an archer, assessing the directions. The lines pass through the air without any obstacle when the measurement is subject to many parameters. I was wrong, many historical facts changed my mind. I thought that my manual experience was closer to an origin (completely fantasized) than my school education. Now, the hand is as cultured and contemporary as the spirit that guides it.

In ancient civilizations, from India to Egypt, the references to the rope take a ritual and religious aspect. The Vedic string tightener (*sulbaka*) as well as the Egyptian *harpedonapte* deploy a knowledge whose origins are confused with those of architecture – a major aspect of Neolithisation. At the other end of the time, the Renaissance, the tarot images pay tribute to the rope : especially in the figure of the Hanged Man, the XIIth blade, a symbol of initiation and study. It is necessary to rely to the obvious: the angles of the grid are translated on the ground with a logic of grid, according to the measurement.

The concepts of area and volume are useful for the practical realization of a building : they enter into the evaluation of the weight of the materials, as writing allows to administer the site. This part of the work involves margins of error, if only for safety reasons. One does not skimp on the ropes when it comes to tow a stone of tens of tons on the ramp of a pyramid.

2 – Manifestations of φ in the triangle 3-4-5

1 – By measurement

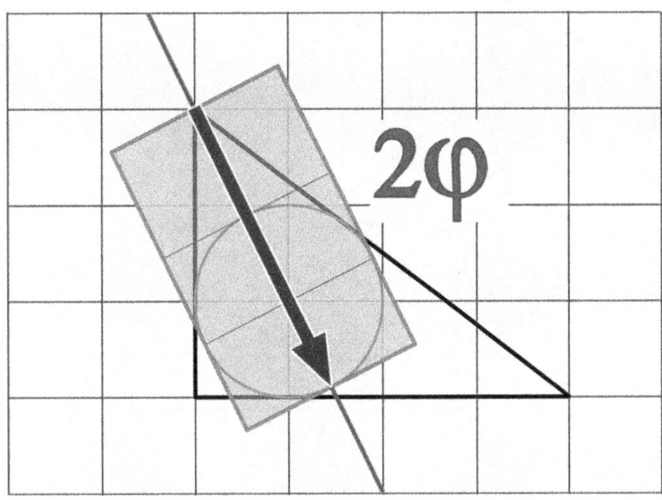

Fig. 43 : The measure of 2.φ on the bisector of order 2

1st manifestation : the golden rectangle 2 x 2φ at its place, called "classic", riding on the bisector of order 2. 2φ is the distance between the apex of the triangle and its inscribed circle.

2 – By the diagonals (angles)

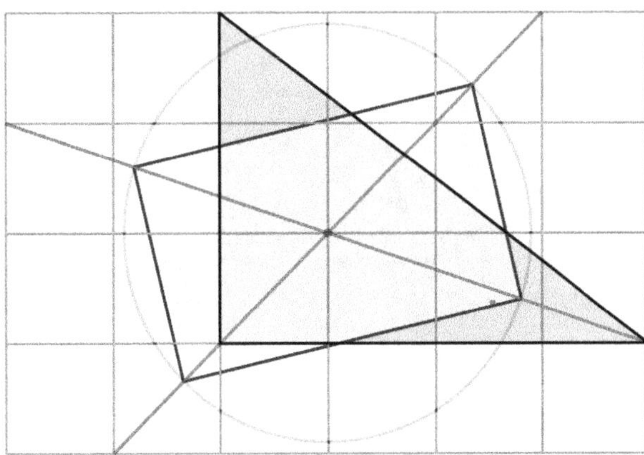

Fig. 44 : The bisectors become golden diagonals

2nd manifestation : The bisectors of order 1 and 3 are the natural diagonals of a golden rectangle. The first manifestation concerns the measurement, the second the angles.

3 – By the limit point

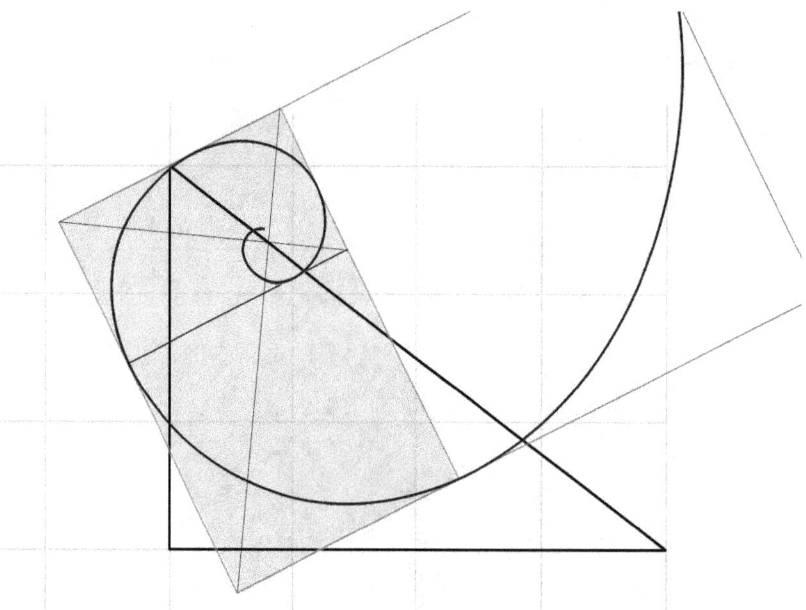

Fig. 45 : The limit of the golden spiral is on the hypotenuse

3rd manifestation : the classical rectangle is divided into an unlimited number of kites (which are in fact disguised golden rectangles). These forms build a golden spiral that converges to the point T of the hypotenuse, located one unit from the top of the triangle.

4 – By the golden area

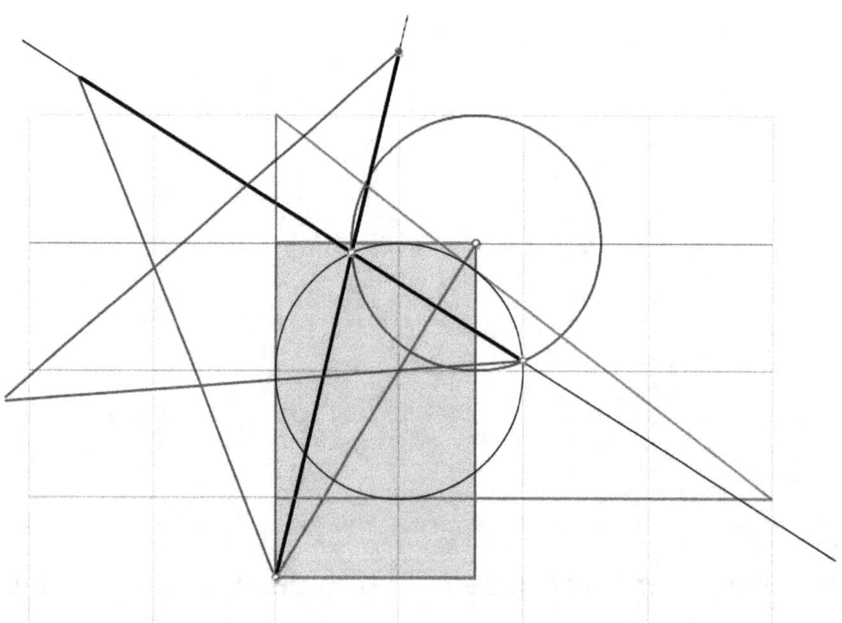

Fig. 46 : The area of $\varphi \times \varphi^2$ leads to a pentagram

4th manifestation : a pentagram comes from the confrontation of the inscribed circle with its twin. This second circle puts its center two unities above the base of the triangle (side 4), and at a distance of φ from the side 3. This pentagram is inclined like the diagonal of a golden rectangle.

This exceptional figure can be simply stated. It is enough to glue a rectangle of φ x φ² on the side 3 of the triangle 3-4-5, at height 2. The upper right corner becomes the center of a circle of radius 1. It cuts its twin, the circle inscribed To the triangle, into two points that decide a pentagram. Under these conditions, the pentagram finds at the bottom the left corner of the golden rectangle, and on the right the center of the inscribed circle !

Excepted the golden spiral, whose point of impact shows the unity on the hypotenuse, the other three expressions of the golden number pass through the center of the circle inscribed on the triangle. This shows the major importance of the structure of a geometrical figure, in comparison to its ordinary borders.

We are going to show these four properties of the golden ratio that history, painting and mathematics had forgotten.

3 – The golden measure in the sacred triangle

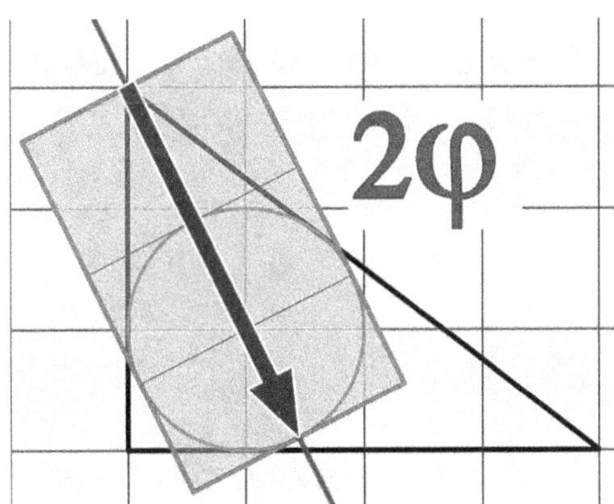

Fig. 47 : The golden ratio in the triangle 3-4-5 (cf. Fig . 43)

The monstration of this property is already done at Chapter IV, 2 devoted to the golden ratio. The length of this rectangle is, according to the convention established on this occasion (∂ diagonal of a double square) : ∂ + 1 . This rectangle is called "classic".

Correlates (secondary properties)

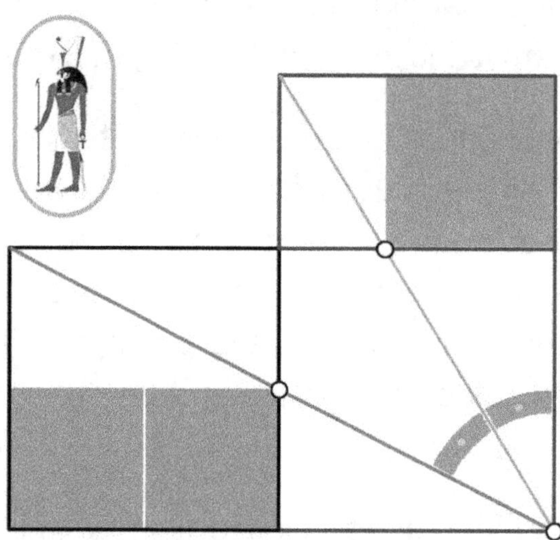

Fig. 48 : The corollary properties of the golden ratio

Golden ratio

When the diagonal of a golden rectangle passes its inscribed square, it puts on the wider side a segment equal to the remaining height to go up to the corner.

Double-square

When it passes the first square, the diagonal of a double horizontal square leaves on the side a segment equal to half the horizontal distance which it still has to cross to reach the angle.

4 – Golden diagonals

We will show that the bisectors of order 1 and 3 of the triangle 3-4-5 are the natural diagonals of a golden rectangle. The golden ratio appears for the second time in the internal structure of the sacred triangle.

Observation n°1

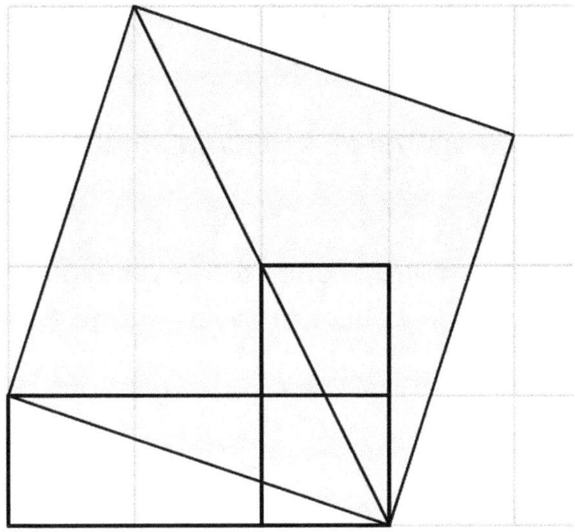

Fig. 49 : Relations between bisectors / diagonals

The figure of the three diagonals

This figure expresses a kind of "principle of the Trinity" : *The angle between the diagonal of an horizontal triple-square and that of a vertical double-square is the angle of the diagonal of a simple-square.*

Observation n°2

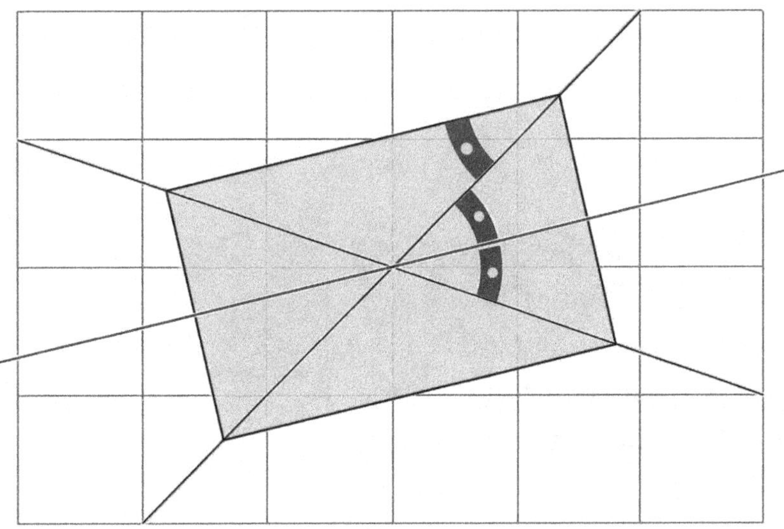

Fig. 50 : The bisectors of order 1 and 3 of the triangle

We know that the small angle of the diagonal of a golden rectangle is half the wide angle of the diagonal of a double square. According to which :

The small angle between the two diagonals of a golden rectangle, in its center, is equal to the wide angle of the diagonal of a double square.

Correlate - Property

When a golden diagonal is horizontal, the other is that of a double square.

This is, as we mentioned, one of the principles which will be used by Dürer in the construction of his engraving « MELENCOLIA § I ».

Monstration

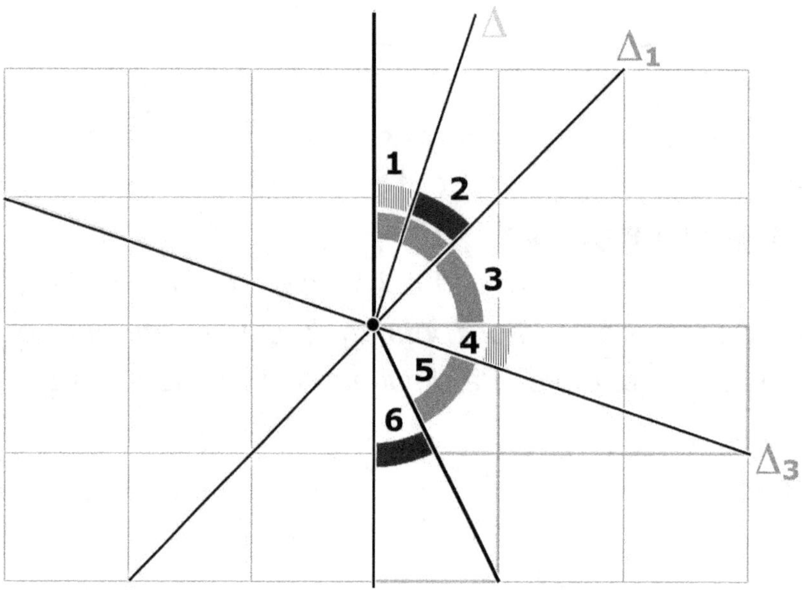

Fig. 51 : The counting of angles

We will note the angles as $\hat{A}n$
Eg : $\hat{A}_3$ for the angle n°3
According to the preamble N° 1 : $\hat{A}_5 = 45° = 90° \div 2$
Let Δ the perpendicular to Δ_3
Δ_3 is the diagonal of a triple-square
—> The angle $\hat{A}_1$ is thus equal to $\hat{A}_4$
We also note that, by construction:
$$\hat{A}_1 + \hat{A}_2 = \hat{A}_3 = \hat{A}_5$$
—> Therefore : $\hat{A}_3 + \hat{A}_4 = \hat{A}_5 + \hat{A}_4$
Now: $\hat{A}_5 + \hat{A}_4 =$ Wide angle of the diagonal of the double square,
—> Therefore $= \hat{A}_3 + \hat{A}_4$ too.   *What Was to be Shown*

Correlates - Properties

*The bisectors of order 1 and 3 of the 3-4-5 triangle
are the natural diagonals of a golden rectangle*

The small angle formed by the diagonals of a simple and a triple-square is equal to the wide angle of the diagonal of a double-square.

5 - The golden spiral of the triangle 3-4-5

The golden rectangle we studied with the first expression of the golden ratio holds kites also marked by φ. At first glance, this new structure is applicable to any golden rectangle, but the 3-4-5 triangle particularizes the properties : the hypotenuse of the triangle passes through the precise point (T) where intersect the diagonals of the said rectangle and its residual rectangle, and this is the limit point is the spiral of kites. The golden rectangle does not adopt this "favorite position" only to give consistency (the surface) to the measure 2φ...

"T" - Remarkable point of the triangle and the golden rectangle

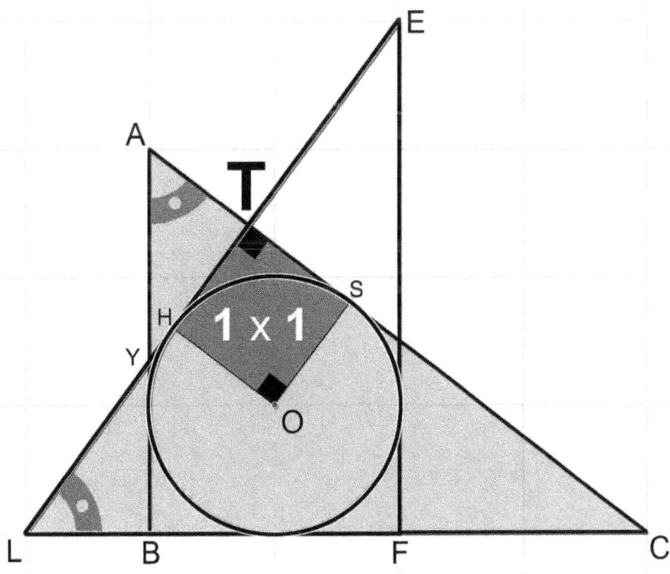

Fig. 52 : The point T of the hypotenuse, at 1 unit of the vertex

Let the triangle LEF, the rotation of ABC by a quarter of a turn, and sharing the same circle. By construction :

LC = AC = 5.

The angles in A and L are the same.

The lines AB and LE intersect symmetrically in Y, relative to the axis YC of the form - here in light grey.
And the center of the inscribed circle is on the line YC.
According to the symmetry of this form :
LB = AT = 1 (explicit on the grid).
A 90° rotation links the triangles.
The hypotenuse LE is orthogonal to the hypotenuse AC.
A square 1 of 1 highlights the points of tangency of the hypotenuses H and S. LE is thus the perpendicular bisector of the segment AC, and T is the middle.   *What Was to be Shown*

Remarks

These achievements will greatly simplify the continuation of this exploration. And we will take the same type of figure to complete its measurements.

Second remarkable point of the golden rectangle (Ω)

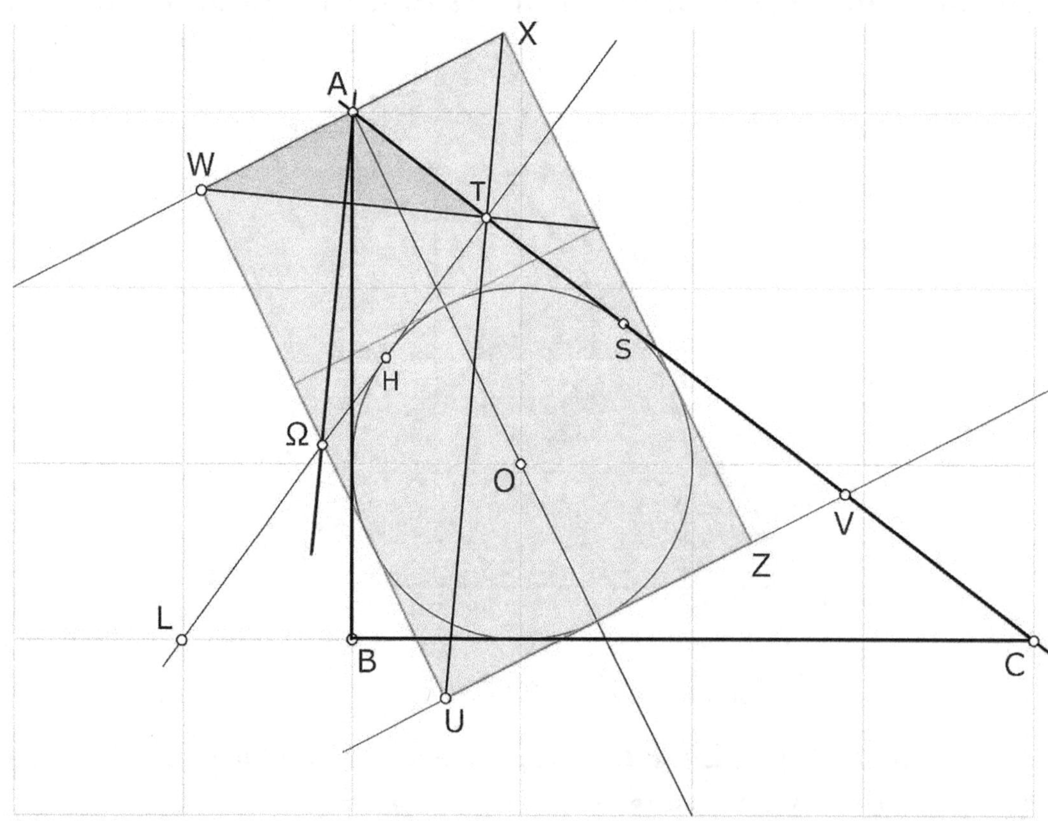

Fig. 53 : The point Ω, tip of a golden kite

Consider the triangle WAT. Call :

$\hat{A}_A$ the angle WAT

$\hat{A}_T$ the angle ATW

$\hat{A}_W$ the angle AWT

α the small angle of the diagonal of a golden rectangle.

—>   2α is thus the great angle of the diagonal of a double-square.

Consider the angle $\hat{A}_A$

On the left of the vertical AB, the angle is 2α,

and on the right : 2 x [90°- 2α]

ie two times the small angle of the double-square

—> $\hat{A}_A = 180 - 2α$

We have shown that $A_T = 1$

Now WA = 1,   therefore WAT is isosceles, therefore $\hat{A}_W = \hat{A}_T$

The sum of the angles of a triangle is 180° (or π).

We can borrow this property.

$\hat{A}_W = \hat{A}_T = [180 - \hat{A}_A] \div 2 = [180 - (180 - 2α)] \div 2 = α$

Thus T is on the diagonal of the residual golden rectangle.

Let AΩ the perpendicular bisector of WT (diagonal of the residual golden rectangle).

Let XU the diagonal of the golden rectangle.

XU is also orthogonal to WT

The orthogonality of the diagonals is one definition of the golden ratio. In fact :

If the residual rectangle (obtained by removing the inscribed square) has the same proportions like the original golden rectangle, its 90° orientation with the first causes the same effect for its diagonal.

—>   AΩ is thus parallel to XU

—>   ΩWA and UWX are thus similar.

By construction, their ratio is 1 of 2 (A is the middle of WX)

—>   T is thus on XU ant WATΩ is a kite

—>   ΩW = ΩT is thus the half of WU, ie φ.

Other remark

The angle in W of ΩWA is right,

therefore the angle in T of ATΩ is right.

—>   H is thus on ΩT (HT is orthogonal to AS)

The spiral of the kites

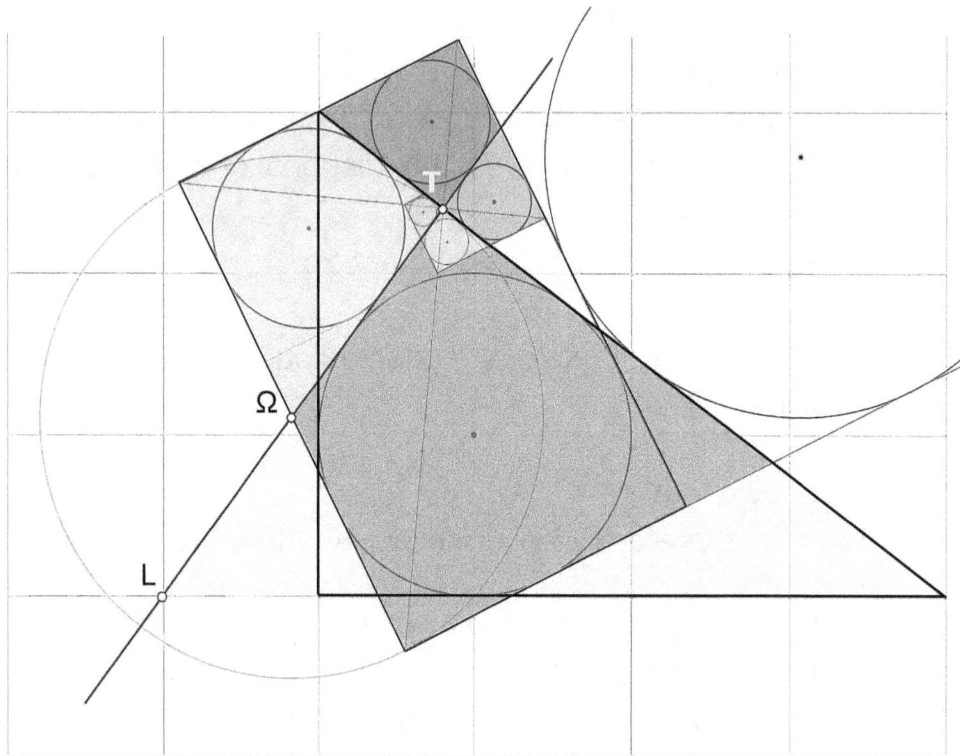

Fig. 54 : The spiral of the kites

Look at the grey kites. We have found the measures of the light grey quadrilateral : 1 and φ. What about the light grey one ? Its "crosstie" is by construction, a segment of the diagonal of the golden rectangle, thus orthogonal to that of residual rectangle. The measure of the two head segments of the quadrilateral is φ (established). The angles at the point T, and at the bottom of the rectangle, complete a kite, similar to the light grey one. Because the angles of the head triangles are the same (last element, α).

The same type of observations on angles and segments, allow to continue the chain of kites. They are glued to each other with a 90° rotation and they reduce the preceding measures by φ. It is interesting to place the incircles at each step. The light gray one is the inscribed circle of the triangle 3-4-5. These circles are tangent to the four sides of each kite, and they serve as an alternative to the conventional presentation of the golden logic, which appear only with the squares.

The kites, fruits of successive divisions by the golden ratio, combine to form a golden spiral. It converges in T, the point where the diagonals of the golden rectangle and its residue intersect   the hypotenuse of the triangle - at a unit from the upper point.

NB : the kites have sides of type (1, φ) joined by a right angle. Both triangular halves can be rearranged to form a golden rectangle. In this case, the series of kites takes another form (See Fig. 56).

Complements to the first figure about the point T

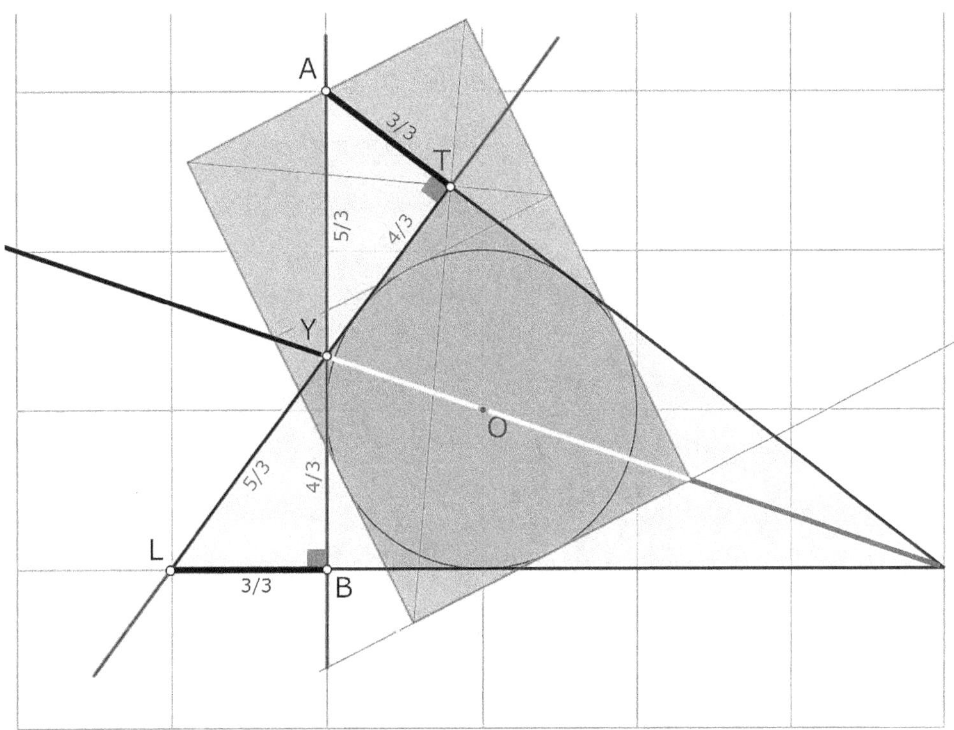

Fig. 55 : How to measure with symmetrical figures

The triangles BYL and TAY are symmetrical on either side of the bisector of order 3 (CO diagonal of a triple-square).

These triangles are right and they share an other angle with triangles of type 3-4-5.

The sum of the three angles is equal to 180°, thus the third angle is the same too. BYL et TAY are of type 3-4-5.

LB = 1. So just divide the measures 3, 4 et 5 by 3 to solve the figure of triangles BYL and TAY.

The results are 3/3, 4/3 and 5/3.

The height of the 3-4-5 triangle, standing on its hypotenuse,
is   (3x4)/5 = 12/5

—>   It is the "altitude" of the point T above LC.

—>   Consequently, T is located at 12/5 x 1/3 = 4/5
on the right of the vertical AB.

If we stick to its classic golden construction the point T has an irrational status. This figure explains, by the magic of angles, that its position on the grid is perfectly rational.

This comment has to enter the file of the "crisis of immeasurable" that shook mathematical Greece.

The golden rectangles

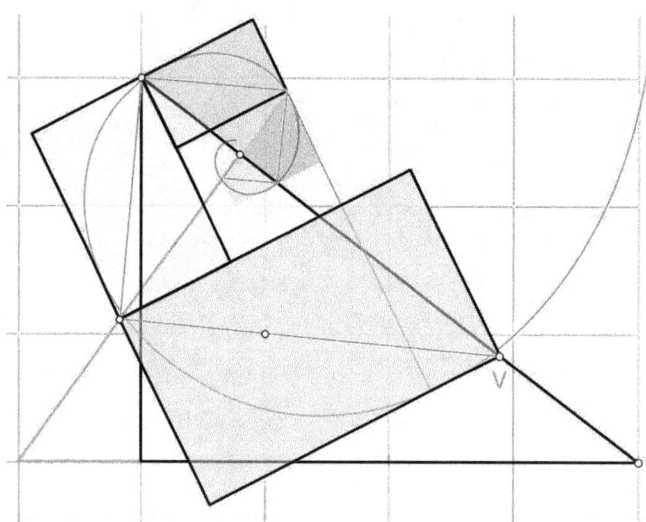

Fig. 56 : Golden Rectangles – Reconstituted from Kites

In this set, the measurements take for each that of the previous rectangle and adds its golden reduction at right angle. The figure is a true subject of meditation.

The golden spiral

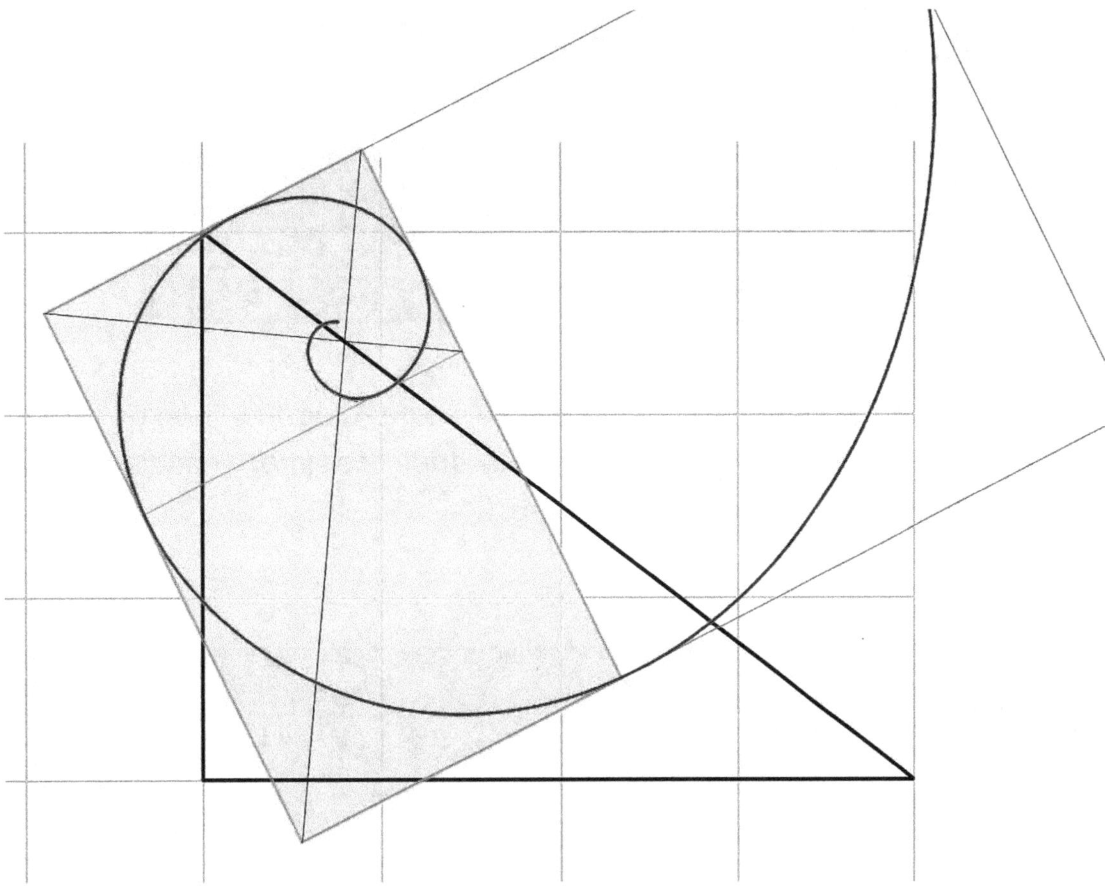

Fig. 57 : The golden spiral of the triangle 3-4-5

As abstract as appears this curve, it is part of the internal structure of the triangle 3-4-5. As such, it contributes to a symbolism which is based on the geometry as it is demonstrated mathematically.

In artistic composition, the shapes guide the representation. In this exercise, all the theoretical arguments are put to use. As the "geometers" discover the structures, they find in them (and not give to them) an inherently logical sense which is the basis of their design.

The relationship of the point T with the inscribed circle

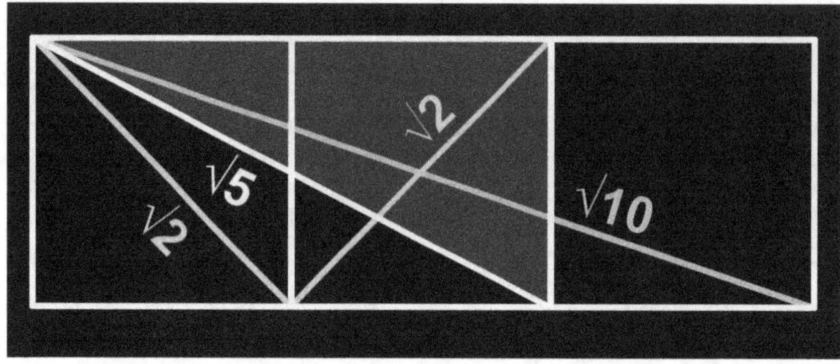

Fig. 57 b : The triple square and its diagonals.

Consider a triple square: 1°) The large diagonal noted √10 in modern way, crosses that noted √2 in their common middle. 2°) A pinwheel triangle is distinguished 2-1-√5.

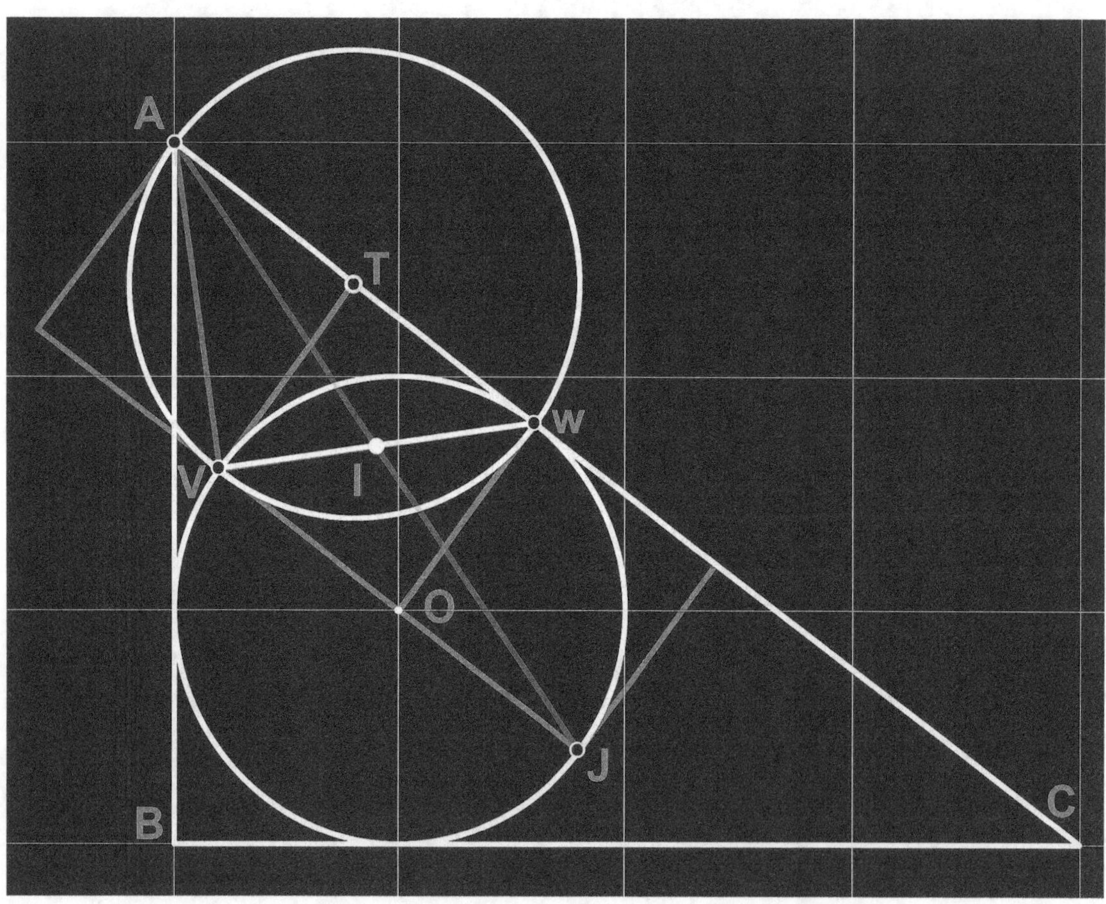

Fig. 57 c : The circle of radius 1 at point T

Let us place the preceding figure on the hypotenuse, according to A and T, and draw the circle of radius 1 at the point T. It passes through A and W. Then AOW is a pinwheel triangle with its right angle in W. This point is therefore the orthogonal projection of O onto the hypotenuse, and the point of tangency of the inscribed circle.

OW = TW = 1, TV = OV = 1.   VW is the width of the almond.

The double square, inscribed

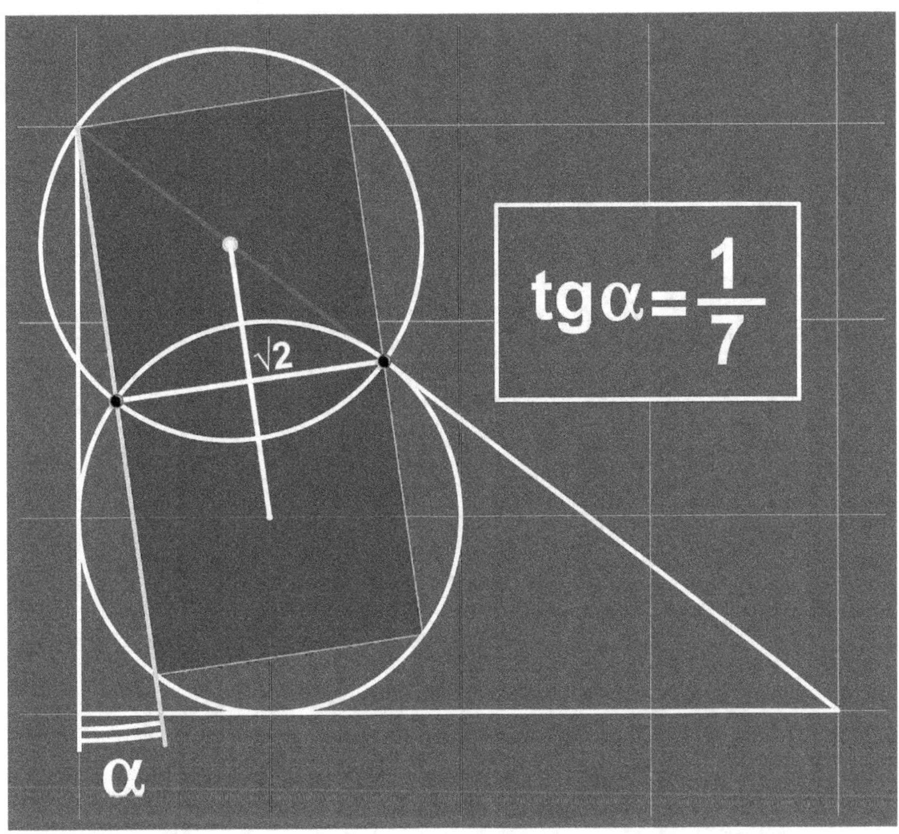

Fig. 57 d : The double square of the circle at T

WVA sketches a square of side √2, inscribed into the circle - it is the width of the almond - intersection of the circles. One can then complete the figure of the double-square inscribed in the double-circle. The angle of this rectangle according to the vertical has a tangent of 1/7 (next page, Fig. 57 e). This explanation was lacking : unlike the other three manifestations of the golden ratio, the golden spiral - with its limit at the point T - was not directly related to the inscribed circle of the triangle.

The interpretation of this cross of two √2 raises questions. The pure value of √2 is an object of caution among the Egyptians, and participates in representations of the devil by Albrecht Dürer. In music, in the equal temperament range, the augmented fourth (do-fa♯) and the diminished fifth (do-solb) are equal and their measure is six semitones; They have a frequency ratio of √2 with do. This "average" is curiously the process that the Egyptians chose in Giza to expose the value √2 ! The Gregorian chant uses this interval, the triton, but at the end of the Middle Ages it is systematically avoided because judged too dissonant. It received the nickname "Diabolus in Musica" !

The angle Alpha

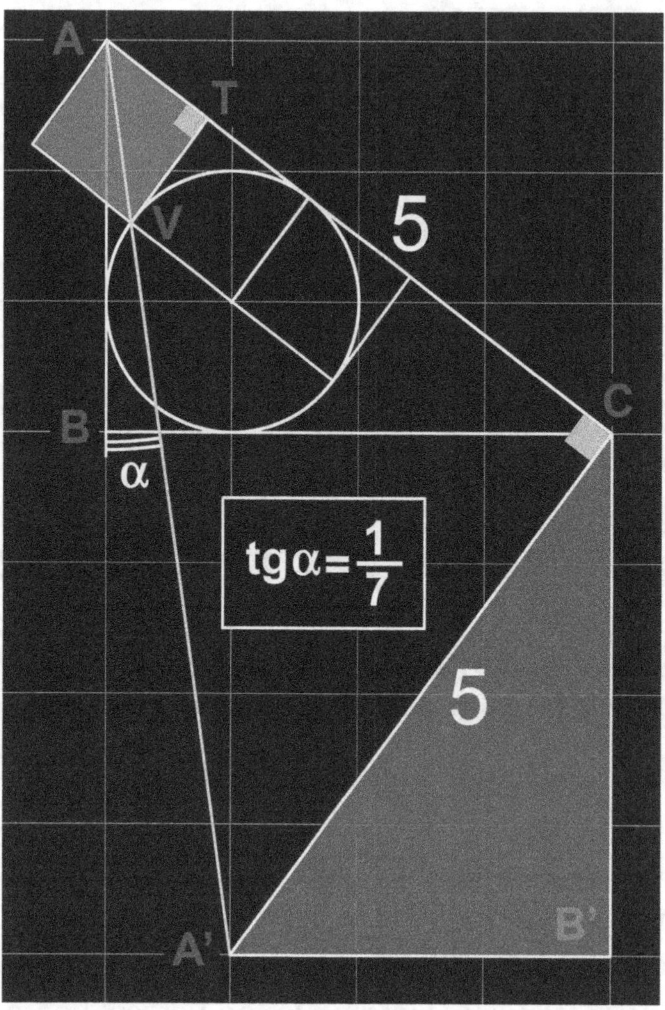

Fig. 57 e : Monstration of the angle Alpha

Let the triangle A'B'C - type 3-4-5.

It is the result of a rotation of the triangle ABC by a right angle around C. Therefore A'C is orthogonal to AC

The triangle ACA 'is a half square, similar to the small half-square ATV, and they share the same hypotenuse.

This line is the diagonal of a sevenfold-square

58

The main composition of the tarots is a series of astrological triangles, type 3-4-5, linked by an homothety (the triangle of the Sun serves as a reference).

These triangles and their grids are turned of 1/4 of a flat angle - ie 45°, and the hypotenuse also becomes the diagonal of a sevenfold-square, as AA'.

6 – The pentagram of the triangle 3–4–5

This time we will see geometry in a modern way, as it seems difficult to develop this chapter entirely without the theorem of Pythagoras (actually the theorem of diagonal).

The DNA of the pentagram

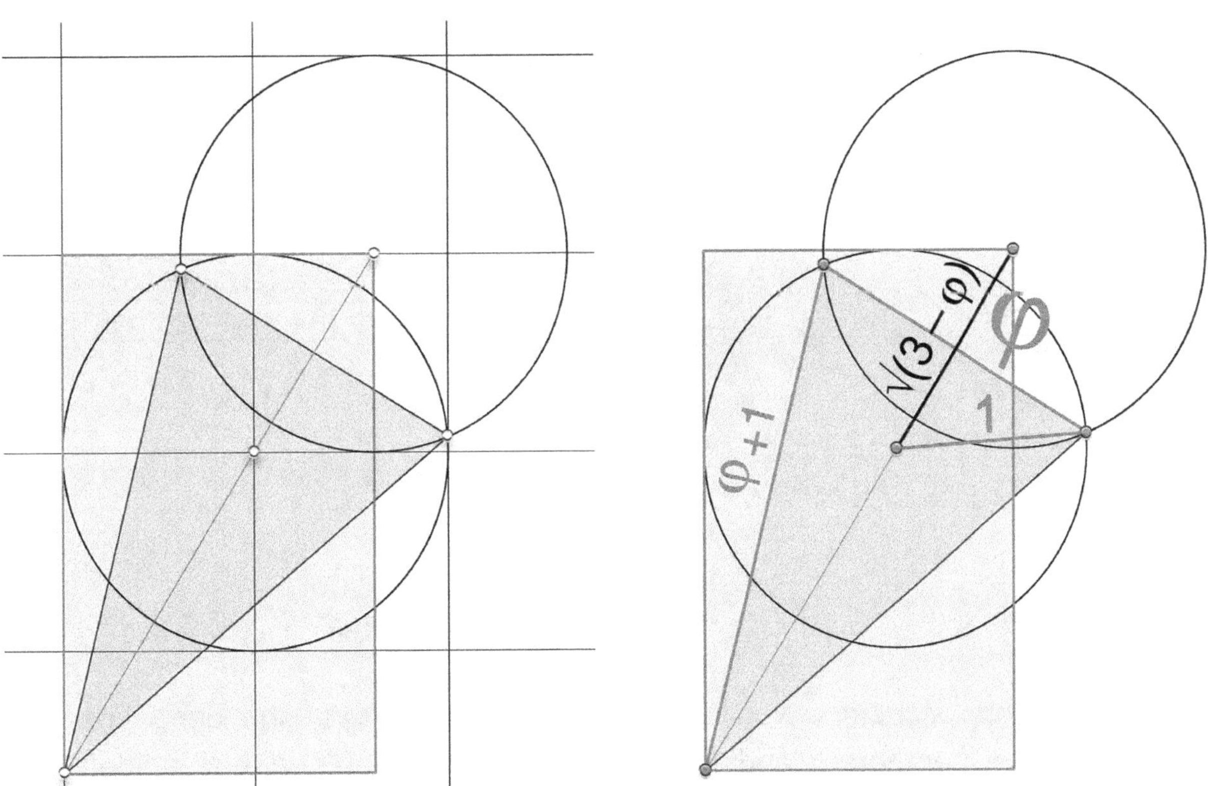

Fig. 58 : The circle with radius φ, at altitude 3

Considering a circle of radius 1, there is only one way to draw a golden triangle $φ^2$ - φ - $φ^2$ of which **1°)** φ is the chord of the circle, and **2°)** the tip is on the vertical tangent of the circle - like in this visual.

Property 1 : *The twin circle which crosses the first on the segment φ is located at the corner of a rectangle of φ on φ² (= φ +1).*

Property 2 : *The segment that joins the center of one of the circles (r = 1) makes an angle of 36° with the axis of the almond.*
> tg 36° = √ [(3 −φ)/ φ]

The pentagram of the triangle of Isis

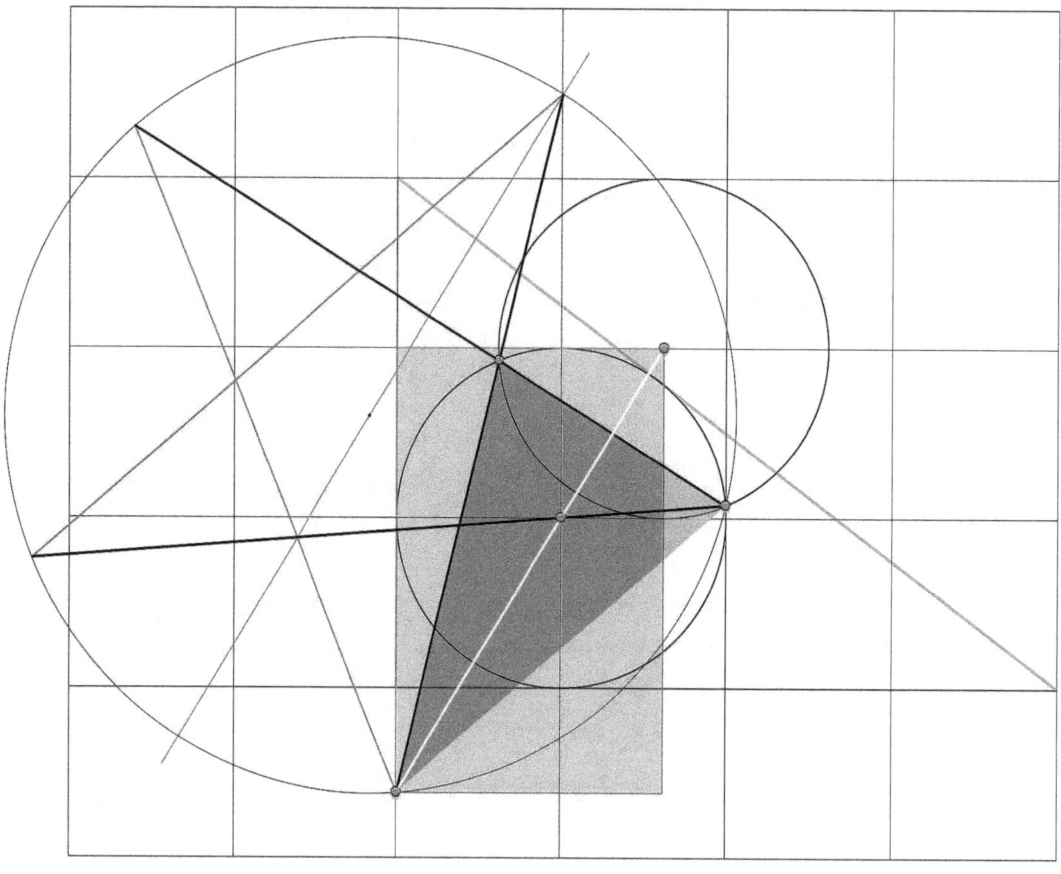

Fig 59 (cf. Fig. 46) : The pentagram of the triangle of Isis

On this occasion, we can once again understand the deep ties between the pentagram and the golden ratio. Golden triangle, golden diagonal, golden rectangle, all these arguments enter the composition.

About the theorem of the diagonal (Pythagoras), it remains unproven that the Egyptians ignored this part of the Sumerian knowledge -- and Babylonian. Above all, they developed geometry in favor of an art devoted to symbolism and the highest aesthetics.

Beginning of monstration

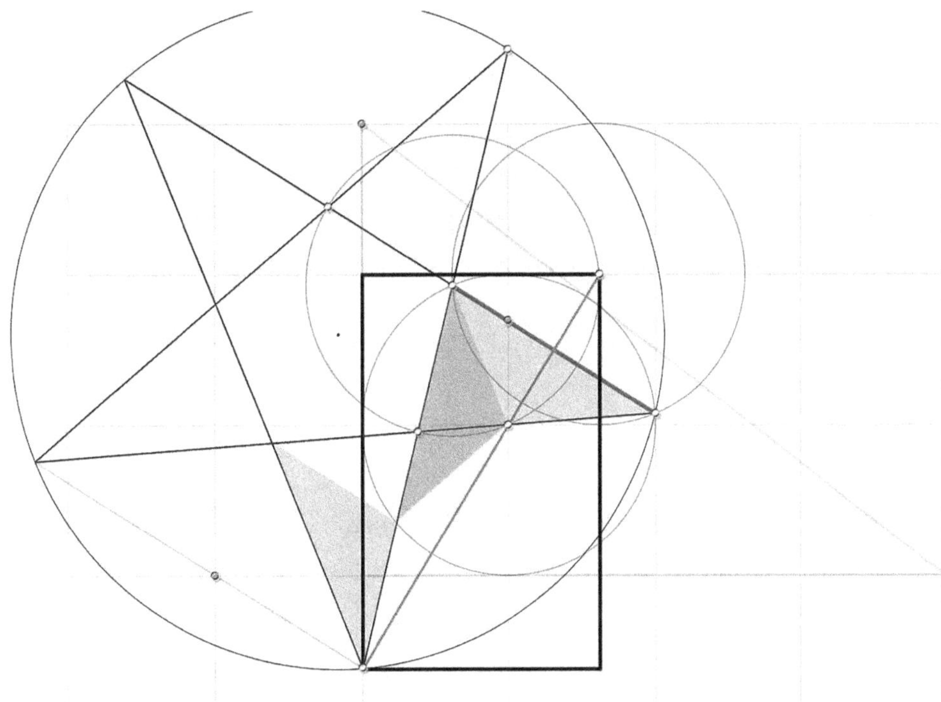

Fig. 60 : The golden rectangle

This pentagram sits on the almond, and φ is the measure of its tip. The star then takes this measure where it originally is, as the top side of a golden rectangle. The pentagram seems to be inclined in an unconventional way, whereas its orientation makes sense: the straight line considered as its vertical when it is in a standing position agrees with the diagonal of a golden rectangle.

The reference rectangle takes the first measurement of φ, which leads to the center of the second circle, and develops downwards with the height $φ^2 = φ + 1$. The Penrose modulus (silver triangles) make this construction particularly didactic.

Remark 1 : The trigonometry reveals coincidences of an accuracy of a few thousandths (margin). On the other hand, those which are accessible to the logic of the eyes are perfectly accurate.

Remark 2 : The practice of sacred geometry developed considerably at Byzantium, where monks were the only ones in Europe to read the Greek in the integral*. There is no doubt that they have reinforced the practice by calculation.

Ref. * : *« Le matin des mathématiciens » (biblio)*

VI – THE SQUARE OF THE POPE

This chapter is partly borrowed from the French article :

« La naissance de la Géométrie : la géométrie avec les yeux des Egyptiens » – revue Repères-IREM N°87 (March 2012)

http://www.univ-irem.fr/spip.php?article=71&id_numero=87

This square takes its name from the circle of diameter $2.\varphi$ that it accompanies. The circle of the Pope has a great importance in the composition of Tarot (version Dürer / Conver). It perfectly describes the curve's of the Pontiff's garment, exposed by the blade V of the Major Arcana. More broadly, this measure $2.\varphi$ comes back often in sacred geometry. It is often the "dimension of the scene" in a painting.

This study reinforces the status of the golden rectangle of the triangle 3-4-5, "classic", and more generally that of the internal structure.

First step

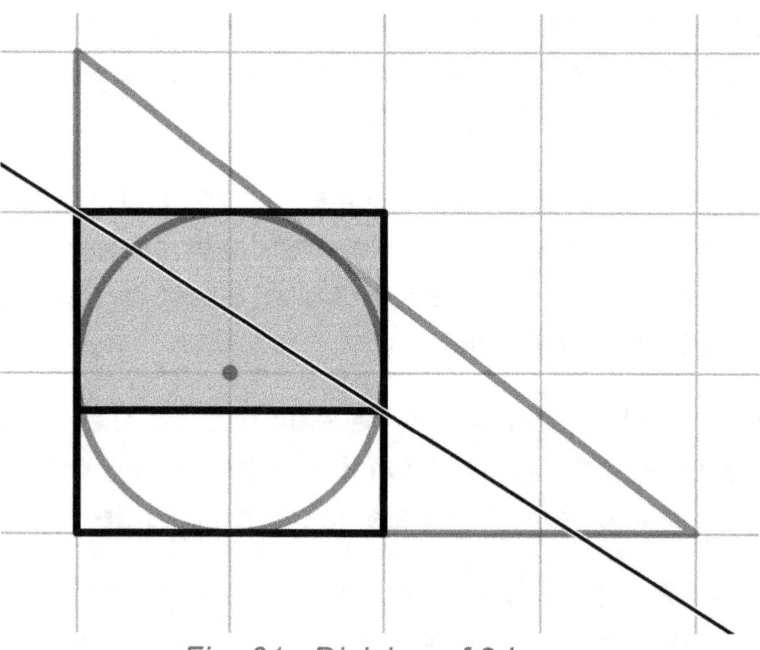

Fig. 61 : Division of 2 by φ

The first step is to divide the square of the inner circle by φ, and to draw a diagonal of the rectangle (which measures 2 by $2/\varphi$).

Second step

A rectangle of width 2, as the incircle, and 2.φ height, stands astride the golden bisector coming from the vertex of the triangle, and takes the first diagonal as its own.

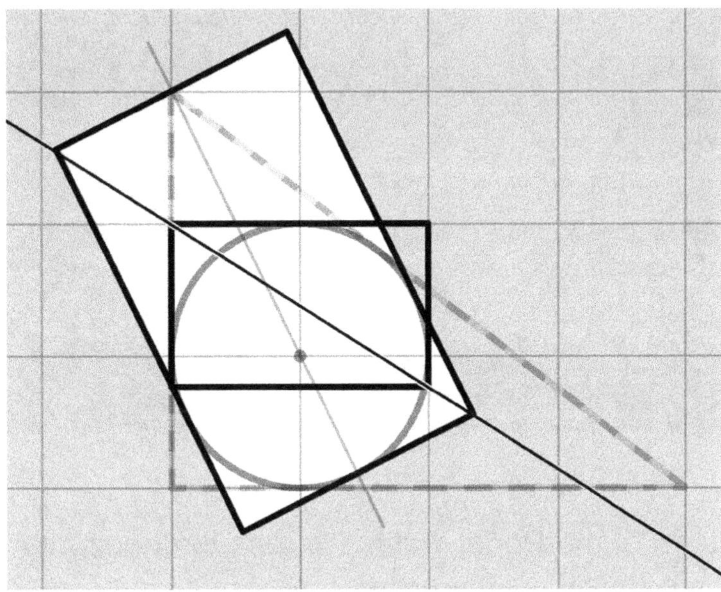

Fig. 62 : A singular property of golden diagonals

Third step

Consider the small golden rectangle inscribed in the   inclined rectangle. Its diagonal intersects the first at right angles, characteristic of golden divisions, at a new golden point (the other is T, that we studied). This point is also on the segment 3 of the triangle, at the height 2, and at 1 unit of the summit.

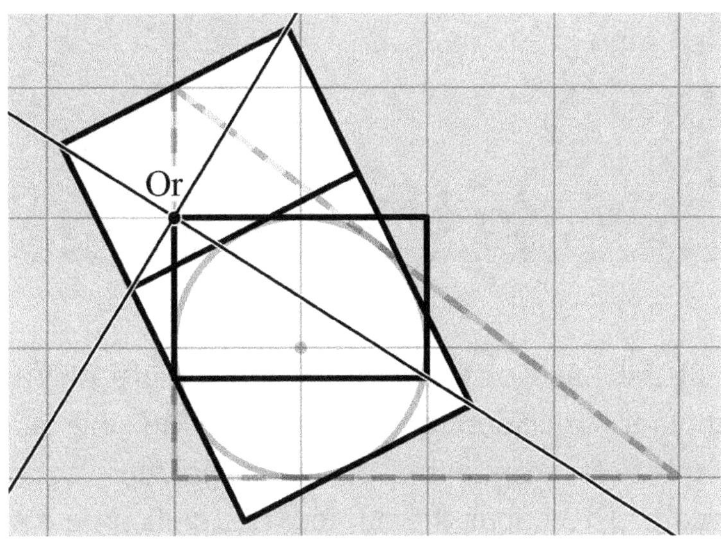

Fig. 63 : The golden point

Fourth step

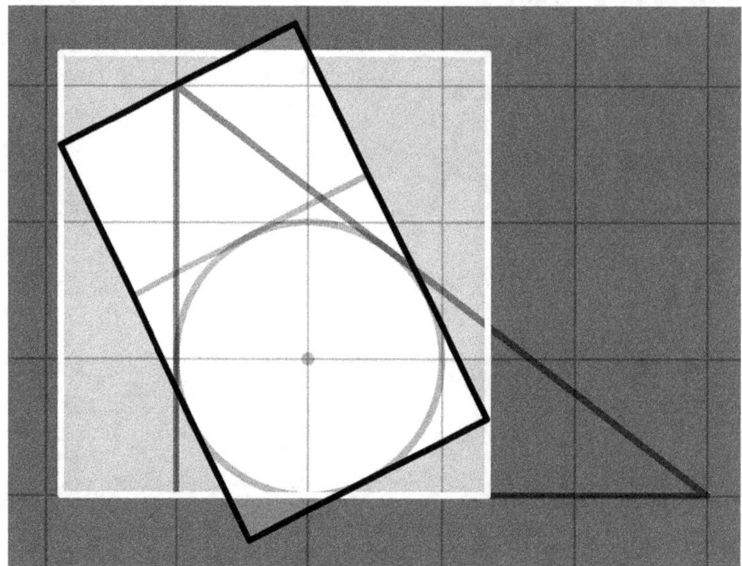

Fig. 64 : The square of the Pope

The square of the Pope, width 2.φ, has the horizontal width of the tilted rectangle. Coincidences multiply.

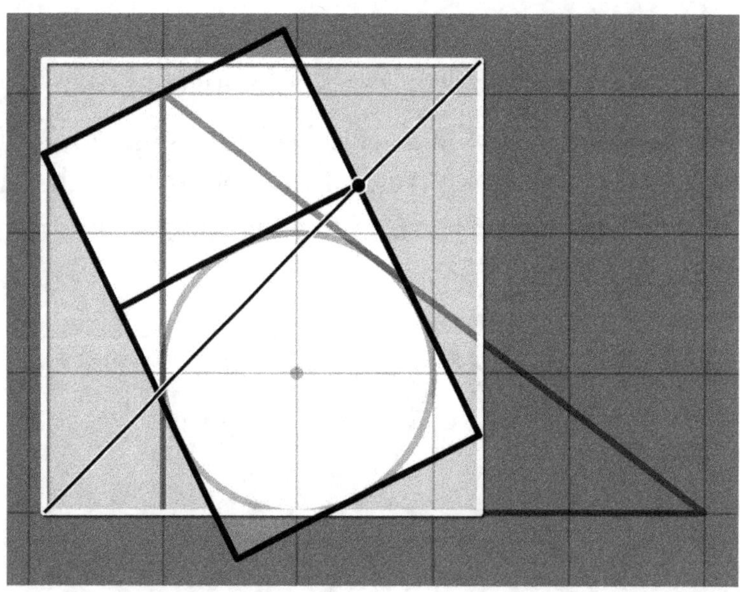

Fig. 64 bis : The diagonal of the square

Placed at the base of the triangle, the square of the pope is the subject of another coincidence: the diagonal of the square passes through the point of separation of small golden rectangle and its associated square. This brings the number of significant points to three.

Fifth step

Finally, the circle of the Pope, of radius φ (same width as the square), arises on the axis of heaven accurately. Its center is on the line between the small golden rectangle and its square. The distance to the edge of the rectangle is φ, the distance to the Earth also.

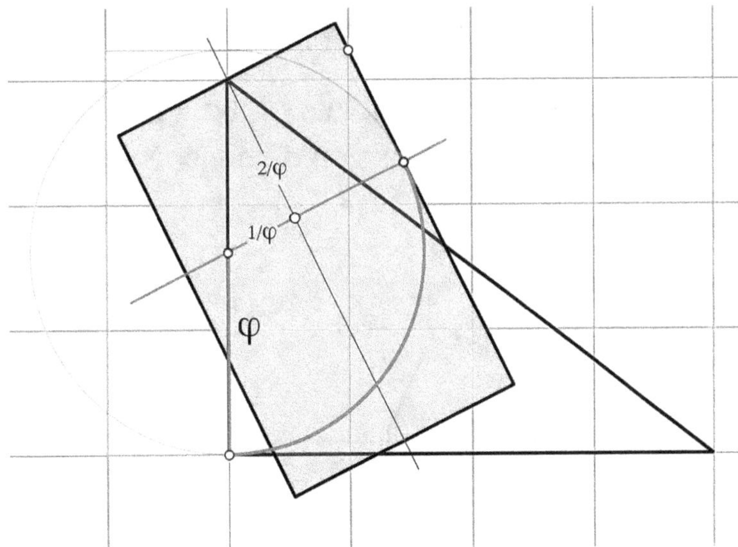

Fig. 65 : Golden arc

Sixth step

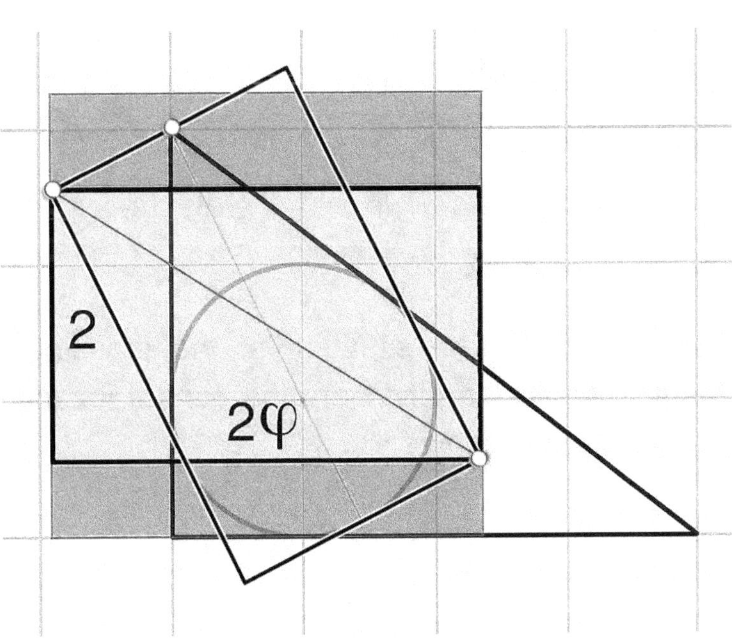

Fig. 66 : The Siamese rectangles

A rectangle called Siamese crosses the first, in a perfectly horizontal position. They share the same diagonal. The definition "with the eyes" of the golden ratio is sufficient to demonstrate this.

Let us start from the common point, bottom right. The long line shares the wide angle of the diagonal of a double square. It is a parallel to the bisector of the angle at the top of the triangle 3-4-5.

Between the two remarkable points of the figure, a beautiful kite is formed once more , a didactic expression of the iso/symmetrical triangles.

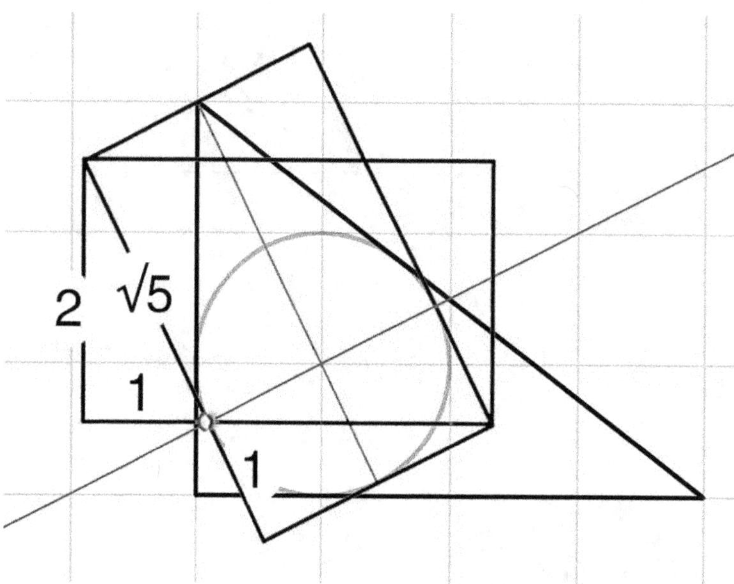

Fig. 66 bis : The measurements

This figure highlights the situation of the crossing point of the Siamese rectangles. One also discovers the central lozenge of side √5, bordered of four pinwheel triangles.

Seventh step

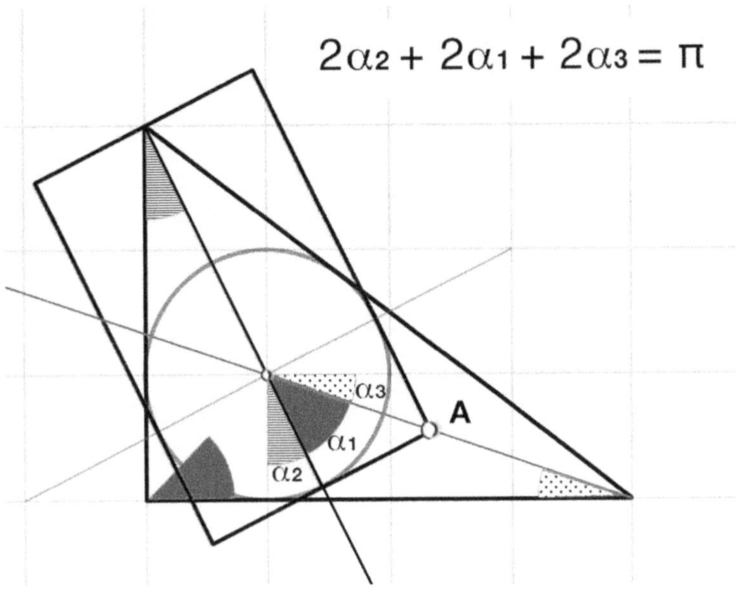

Fig. 67 : The angles

On this occasion, it is realized that the point A, angle of the circumscribed square of the intimus circle, is on the bisector of order 3 of the triangle.

Monstration

It is sufficient to note the sum of the angles of a triangle, and the definition of a bisector to show it. All these considerations underscore the importance and relevance of this golden rectangle - which is called "classic" in the triangle 3-4-5.

Modern Reasoning

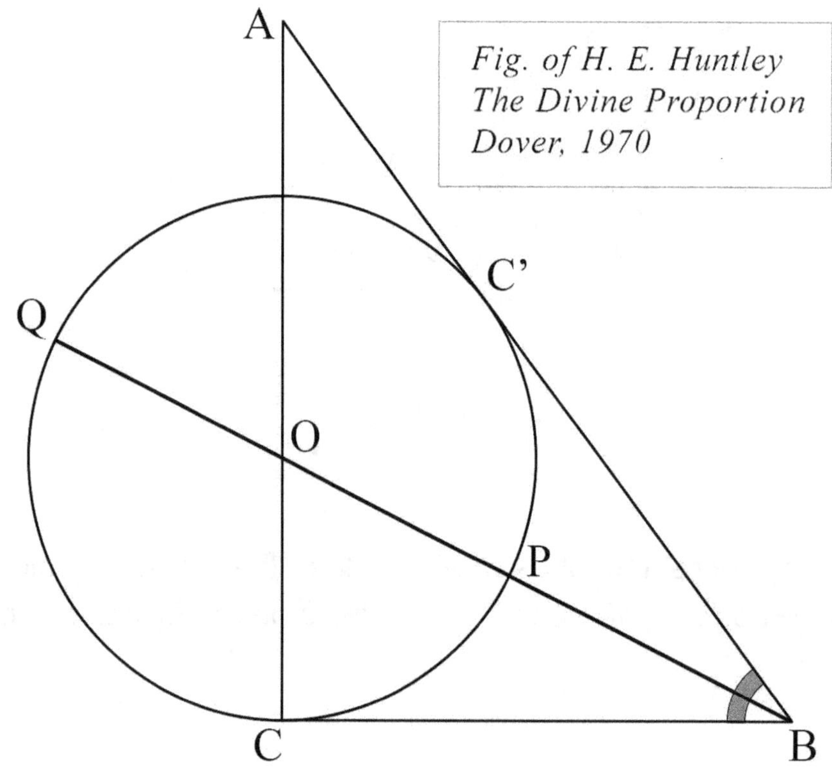

Fig. 68 : H. E. Huntley, The divine proportion, Dover, 1970

Let a circle of diameter 3, constructed at the crossing point of the golden bisector and the *cathetus* 4. HE Huntley goes through the calculation to show that the ratio of this circle and the residue left by its withdrawal on the bisector is equal to φ.   QP/PB = φ

Huntley uses the "power" of a point to show that the circle of 3 is tangent to the hypotenuse. However we have only our eyes to understand. The practice of geometry with the eyes is to the calculation what the bike is to the car - but one does not miss anything of the landscape. In fact, the point of interest is not the ratio of the circle and its residue. The arithmetic value of 3/φ is even a little barbaric. On the other hand there are 3 golden sections. We know the 2φ from the top, and we discover an elegant presentation of its extension of φ.

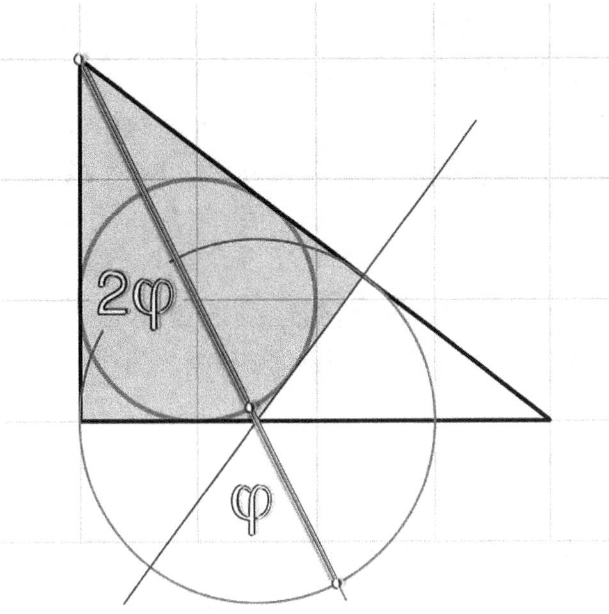

Fig. 69 : The three golden measures of the golden bisector

Pythagorean reasoning

The kite is formed by symmetry of the pinwheel triangle
L = 3/2   H = 3   hypotenuse = (3/2)√5 = 3√5/2
The large circle has for 3/2 radius,
The sum is   3/2 + 3√5/2 = 3 [(1 + √5) / 2] or 3φ

Reasoning with the eyes

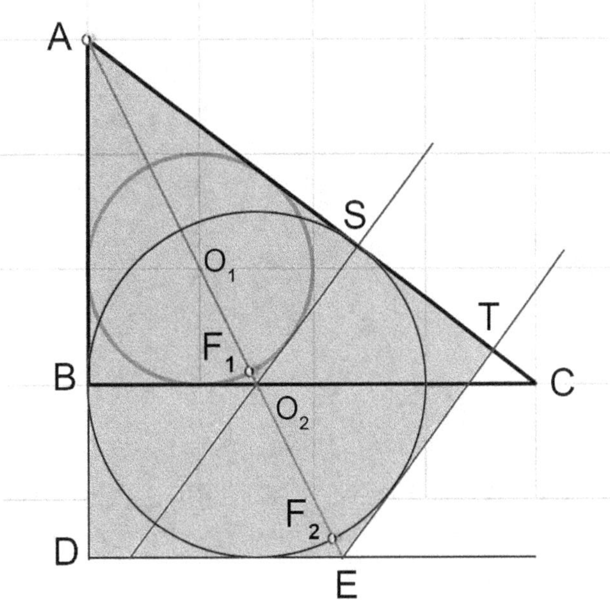

Fig. 70 : The two similar kites

Let the circle of diameter 3 of center O_2
Let the pinwheel triangle ABO_2
Let its symmetric AOS, according to the bisector AE
     The angle is straight at B => right angle at S
—> S is the point of tangency of the circle of center O_2

Let D be the point on the line (AB) such that :
DE, at right angles, is tangent to the large circle

     ABO_2 and ADE have the same angles
—> They are similar, pinwheel triangles
—> ABO_2S and ADET are similar
=> AF_1 / radius of circle of O_1 = AF2 / radius of circle of O_2
AF1 / 1 = 2φ / 1 = AF_2 / (3/2) $AF_2 = 2\varphi . 3/2 = 3\varphi$

What Was to be Shown

The necessity of calculation

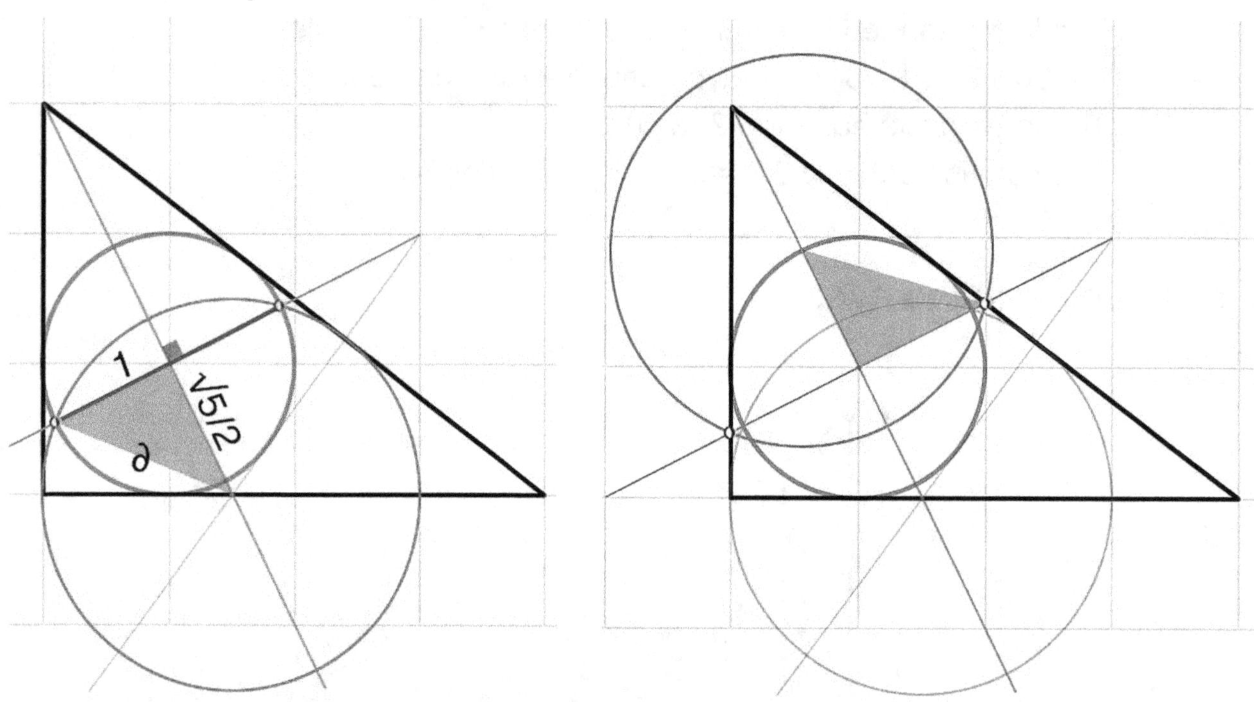

Fig. 71 a et b : The necessity of Pythagoras to solve some problems

On the left, the inscribed circle and the large circle seem to cross each other on the perpendicular - in the center of the first. On the right, the second large circle at the top appears to cross this perpendicular on the triangle. But for now, there is no solution "just with the eyes".

Extension

One can make the parallel between the figures of this list of rectangular triangles : how to build a squared root with the ruler and the compass ?

$$a \qquad \sqrt{(a+b)} \qquad b \qquad \text{with } a < b$$
$$1 \qquad \sqrt{3} \qquad 2$$
$$2 \qquad \sqrt{5} \qquad 3$$
$$3 \qquad \sqrt{7} \qquad 4$$

This is an opportunity to introduce the polynomial formula

$$a \qquad \sqrt{(2a+1)} \qquad a+1$$

that is : $\qquad a^2 + (2a+1) = (a+1)^2$

Additional figure

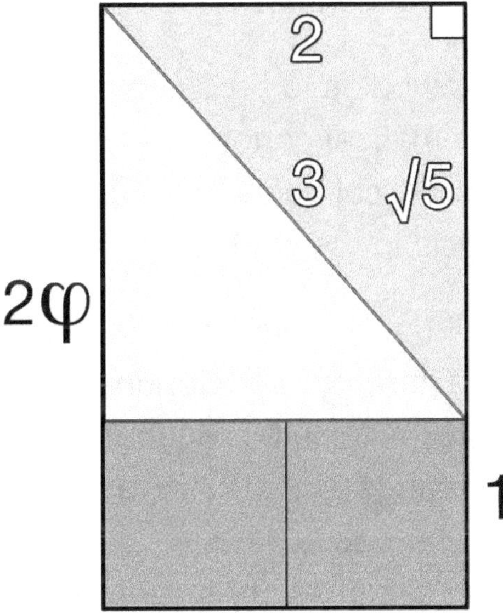

Fig. 72 : Construction by Robert Vincent

The triangle 2-√5-3 allows us to construct simply a golden rectangle of width 2.
http://www.apmep.fr/ADa2-Le-Nombre-d-Or-traces-avec-la

The geometry with the eyes was born before writing, without which the calculation seems impossible. If the state of mind regarding its practice remains the same, it is evident that the sacred geometers have benefited from the contributions of calculation over time, especially the Byzantines, to verify proposals like those presented in this chapter.

VIII – THE LETTER G OF THE TYMPANUM OF CONQUES

We have devoted four chapters to the golden ratio (titled from IV to VII). Would the geometry of the grid revolve around this principle, as many publications invite us to believe ?

For a long time, it was thought that the Earth was inhabited exclusively by men. But we realized that half the men are women. Would it be the same for the figures and principles of so-called sacred geometry ? The magic pirouettes of the golden number have literally occulted the wealth of another kind of value: the root of three.

The purpose of this work is to reconstruct, stone by stone, a corpus of geometry with a mathematical coherence. However, we cannot ignore the projections of this theoretical basis in art and architecture. One learns much about the tool by visiting the works.

This is what we will do by studying a fine example of Romanesque art in the Abbey of Sainte-Foy de Conques. The figure which has mobilized all our attention is a letter engraved in stone. This lapidary inscription (epigraphic) belongs to the Latin text which divides the famous tympanum of Conques. The capital letter G that we are going to study is the first of the Latin word GLORIA. It is part of a set of two lines, here translated into English, where the G is in a way the hinge :

Thus are given to the chosen ones led to the joys of heaven,
Glory, *peace, rest, unending day*

This word is at the foot of Dadon. Of the seigneurial class, and then a self-imposed hermit of Conques, he went on to confront and defeat the Moors in 730. This engraved letter is unique in its kind. Sacred geometry is accustomed to composition marks, but these signs are neutral in nature. They do not participate in any type of discourse, whether narrative, aesthetic or didactic. Their role is to certify the compositions to the reader.

1 – A lesson of geometry

This knowledge is written nowhere else than in the works themselves, and it is the object of great debates. In the present case, not only does the text invades the tympanum, but it also contains an authentic lesson in sacred geometry. This case is an *unicum*.

Fig. 73 : Location of the letter G

In summary, the master of the tympanum asserts that if a square is subtracted from a rectangle of proportion √3, the residue is not a trivial figure: this rectangle has the proportion (1 + √3) / 2. This ratio, very usual in architecture, is obtained by combining the circle with the equilateral triangle. The proportion is here denoted *H*, and it will soon find another.

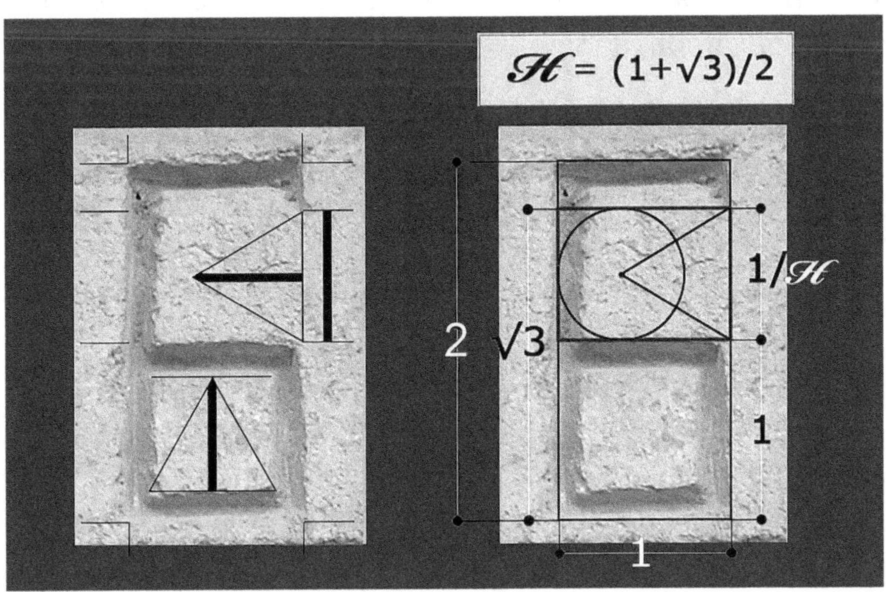

Fig. 74 : The proportions of the epigraphic inscription

A long and careful study precludes the possibility of over-interpretation. All the features here identified as proportions are far too precise. In order to assert the equilateral triangle, the master of the tympanum places in the lower part a false square whose height is that of the triangle (the side of the triangle is explicit in the upper part of the figure). Inspiration and chance would not have this pedagogical talent.

2 – Dialogue of mathematicians with IREM

The "Monstrations of Conques" are a great experience. The pedagogues nourished the composition of their propositions, and the lines of geometry which appeared so familiar gained another status. Like the dialogue between Greece (which left us democracy and pedagogy), and Egypt (where the sacred became art). Is it a coincidence that this exchange has chosen the root of Heaven ?

An appropriate vocabulary

The Egyptians conceive the proportion of the rectangles across their diagonal, and more precisely its angle. Jean-Paul Guichard reminds us that the rectangle of proportion $\sqrt{3}$ chooses the side of an equilateral triangle. To make the thing obvious, it will be called Δ ! Moreover, it reminds us that a rectangle of proportion $\sqrt{3}$ is the assembly of the two halves of an equilateral triangle.

The choice of its geometric echo Δ' [to designate the proportion $(1 + \sqrt{3}) / 2$] integrates algebraic considerations. The series of numbers of their reductions is a kind of "genetic code" :

$$\Delta <\bullet> \sqrt{3} \qquad = [1, 1, 2, 1, 2, \text{etc.}]$$
$$\Delta' <\bullet> (1+\sqrt{3})/2 \qquad = [1, 2, 1, 2, 1, 2 \text{ etc.}]$$
$$\varphi <\bullet> (1+\sqrt{5})/2 \qquad = [1, 1, 1, 1, 1, 1 \text{ etc.}] \qquad \text{for } \textit{information}$$

Crossroads of Properties

Several properties of the square root of three are equivalent, and each can be taken as a definition (by Jean-Paul Guichard).

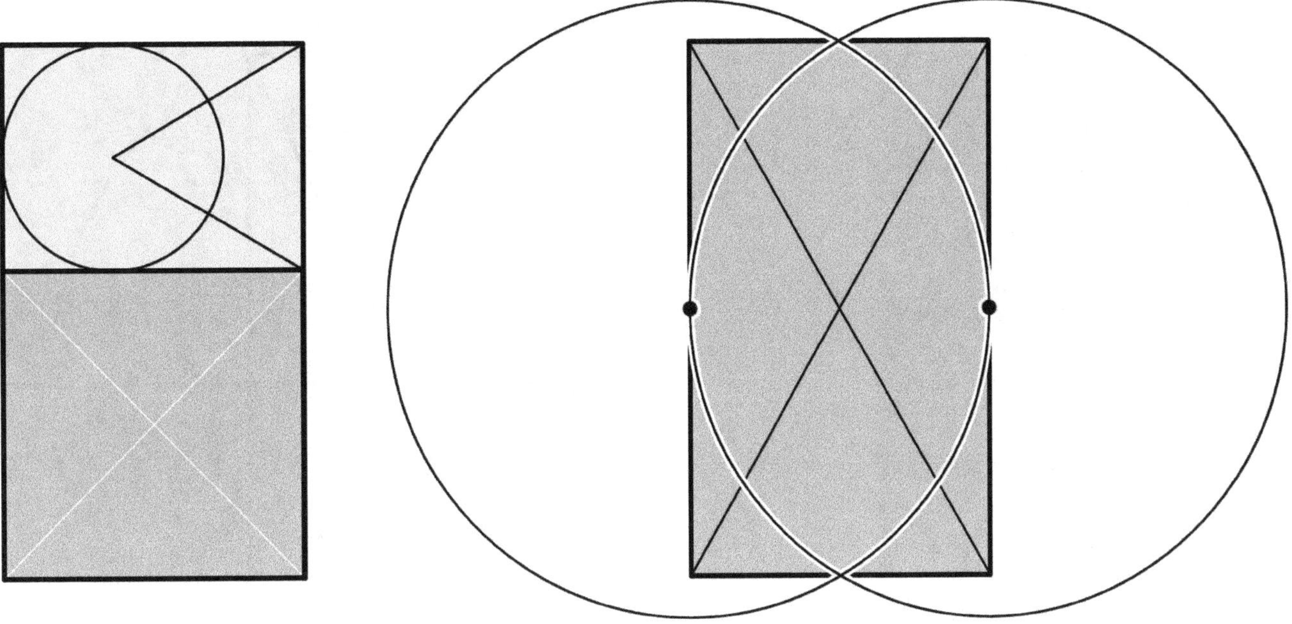

Fig. 75 a : Definition of Conques   Fig. 75 b : Definition with the Vesica Piscis

Three definitions of the root of three as proportion

We can define the proportion √3 as that of a rectangle :

Fig. 75 a – This one, right, built in the style of Conques. A square topped by a figure combining the equilateral triangle and the circle (remember that this figure inscribed in a rectangle type Δ', is a classic of architecture. A lot of manors' doors use it as template).

Fig. 75 b – The rectangle surrounding the almond of the Vesica Piscis. Or a diamond made of two equilateral triangles placed head to tail as a lozenge, or here as a sandglass.

Fig. 76 a – A rectangle with a height twice that of an equilateral triangle, when the side of the triangle becomes the width. Here the ABKC rectangle. This figure is incribed in that of Conques, and this highlights the coincidence of the J point on the diagonal (discussed further property).

Fig. 76 b – The golden ratio also has several potential definitions, but only one can be considered as original : *The small angle of the diagonal of a golden rectangle is half the large angle of the diagonal of a double square.* This definition states the construction process and not just a property (Yvo Jacquier).

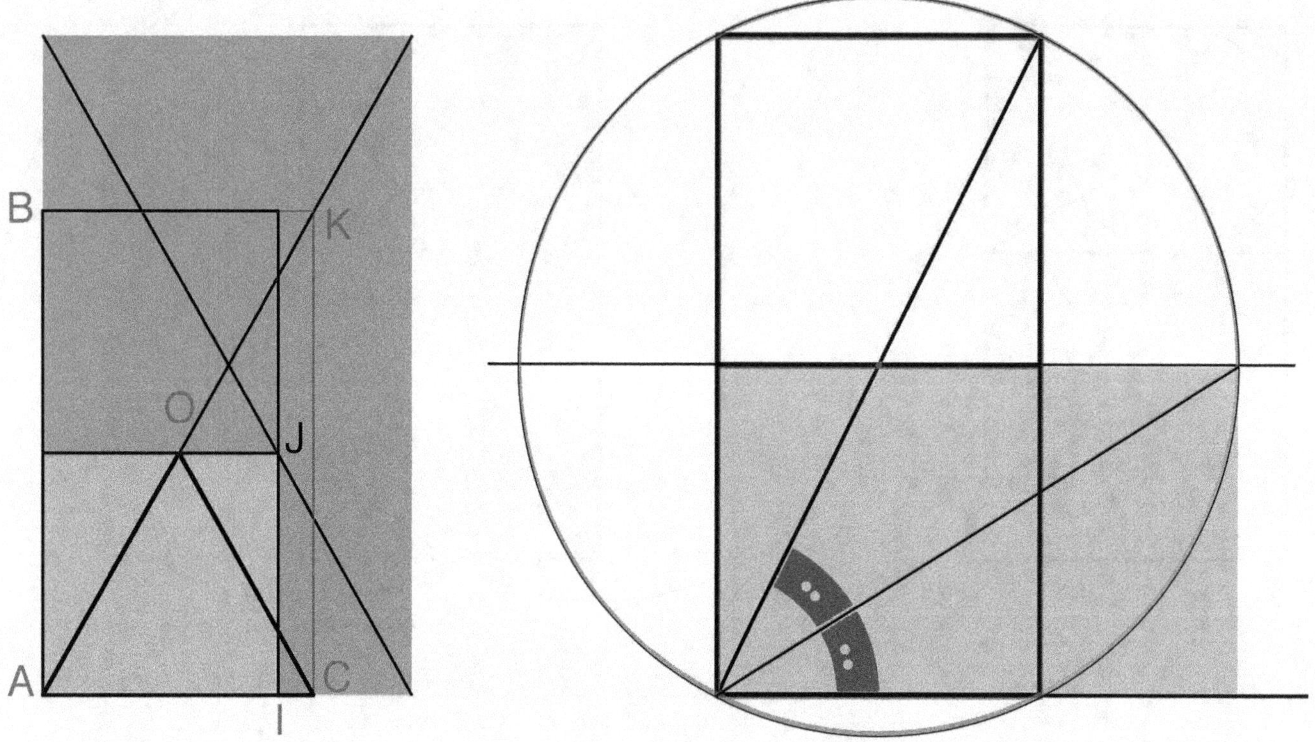

Fig. 76 a : Two equilateral triangles   Fig. 76 b : Construction of golden ratio

3 – Monstrations "like Egyptians"

The first figure

Fig. 77 a – Here is the first figure that emerged in the study of the tympanum of Conques. This « monstration » or demonstration for eyes (without calculation) is the most direct. There remained a great job for teaching, which will allow us to enter the intimacy of the figure.

We see here the dual property that governs the rectangles type Δ and type Δ' :

$$\text{Adding a square to } \Delta' \quad => \Delta$$
$$\text{Removing a square from } \Delta \quad => \Delta'$$

This property is to be compared with that of the golden ratio :

$$\text{Adding a square to } \varphi \quad => \varphi$$
$$\text{Removing a square from } \varphi \quad => \varphi$$

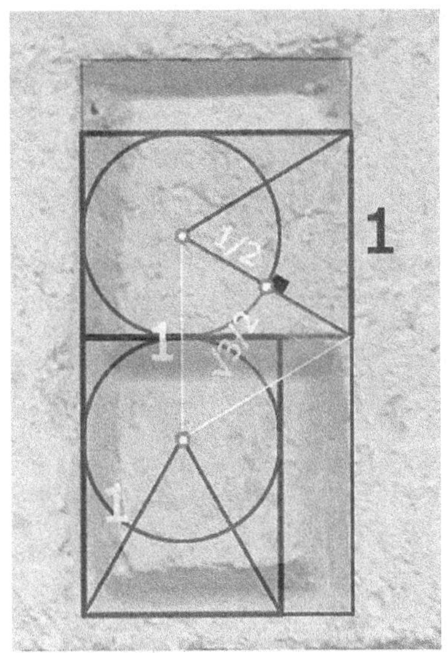 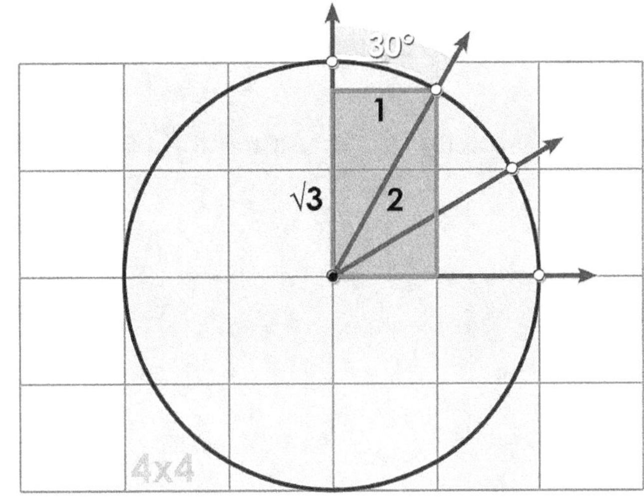

Fig. 77 a : First figure   *Fig. 77 b : Circle of 4 on square of 4x4*

Fig. 77 b – In terms of proportions, the √3 is the result of a calculation type H/L= √3. For the geometry with the eyes, the proportion √3 is the angle of the diagonal of a rectangle of proportion √3. This line is the result of the trisection of the right angle, with 30° on one side and 60° on the other. This trisection is easily obtained with a 4x4 square and a circle of diameter 4.

Monstration of the ratios Δ et Δ'

Fig. 78 a – The rectangle type Δ

The sum of the angles of a triangle is 180°. The equilateral triangle has three angles and 60°. Cut in half at right angle (bisection), we get two right triangles with a peak of 30° and a third angle of 60 °. The rectangle proportion √3, without naming it, can be designed « with the eyes » as the two right triangles head to tail together. A rectangle of such proportion can be called type Δ to evoke its triangle.

Fig. 78 b – Build a square

Build a square of side "1/2 + ∂". Call ∂ the height of an equilateral triangle of side 1. **Use two triangles of sides 1**

It is thus possible to place two triangles continuously.
One at the base, point upward, and the other from its point O, straddling the line of the vertex.

Height of the square   = IO + OJ   = ∂ + ½
Width of the square   = BJ + JC   = ½ + ∂
NB :
The equality of the angles in O shows two right angles at I and J.

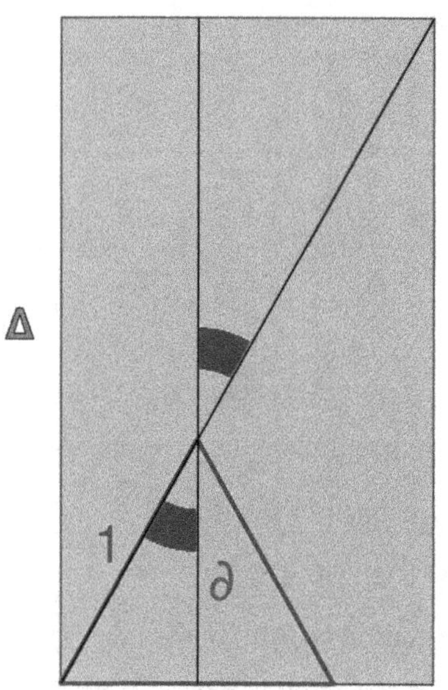

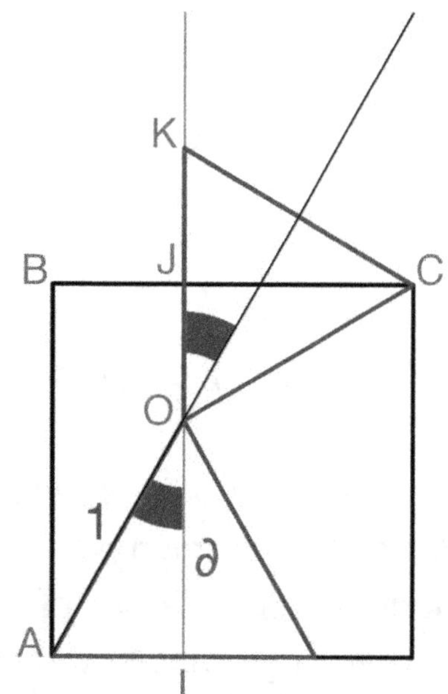

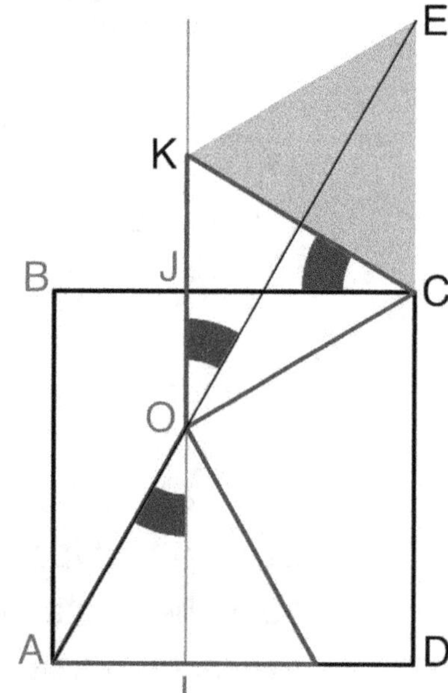

Fig. 78 a : Rectangle △   *Fig. 78 b : Clever square*   *Fig. 78 c : Third triangle*

Fig. 78 c – The third triangle

Build a third equilateral triangle, side 1, stuck to KC
and pointing its peak in E.
The angle $\hat{A}_1$ (= BCK) is 30°
The angle $\hat{A}_2$ (=KCE) is 60°
Therefore
The angle $\hat{A}_3$ (=BCE) is a right angle
D, C and E are (aligned) on the same vertical line

The line starting from A to 60° from horizontal, passes through O
according to the first triangle.
Then, it passes through the middle of KC according to the second
triangle - since it makes an angle of 30° with the vertical. It is thus
the bisector of the third triangle KCE.

The Δ' type rectangle

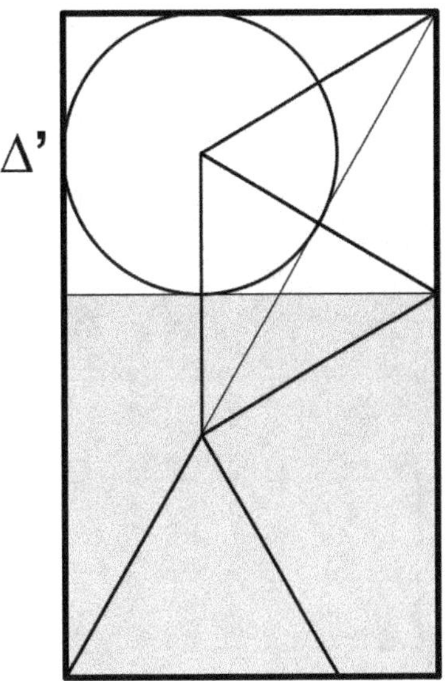

Fig. 79 : The Δ' type rectangle

Now, by its angle, the long line is the diagonal of a rectangle type Δ. We can therefore state that *in any rectangle type Δ, the residue of the removal of a square includes an equilateral triangle completed by a circle of diameter equal to its side.*

NB : The upper part of the figure, the rectangle that overlaps the square is called type Δ', to evoke its link with the type Δ.

The transition to the algebraic reality (Fig. 78 c)

The Pythagorean theorem, more than ever the diagonal theorem, tells us that $\partial = \sqrt{3}/2$ - because the hypotenuse of half triangle is 1, and the small side is 1/2.

According to what the large rectangle has the proportion $\sqrt{3}$, and the residue of removing a square has the ratio $(1+\sqrt{3})/2$.

NB :   The mensurations of the great rectangle are :
Width   AI + ID = BJ + JC = 1/2 + ∂ = $(1+\sqrt{3})/2$
Height   DC + CE = BC + CE = (1/2 + ∂) + 1 = $(1+\sqrt{3})/2 + 1$

If we call $H = (1+\sqrt{3})/2$
$H\sqrt{3} = H + 1$
This equation might be compared to :
$\varphi^2 = \varphi + 1$

Subsidiary figure - 1

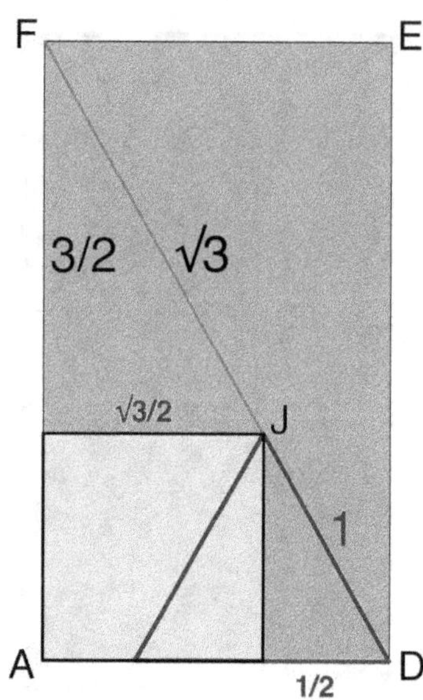

*Fig. 80 : Subsidiary figure **n° 1***

This figure is one of the benefits of exchange, when teaching and research intersect their experience. The third definition of proportion √3 highlights a peculiarity of the figure of Conques. The highlighted square joins the diagonal at the point J.

This is the opportunity to build a didactic figure where two similar triangles meet. Above, the hypotenuse of the triangle is irrational and faces a vertical rational value – √3 and 3/2.

Down we have the reverse version, 1 and √3/2.
The monstration of this figure is in its construction.
Let 1 the side of the equilateral triangle,
as indicated by the master of Conques.
It is known that the figure is width ∂ + 1/2.
[Where ∂ is the height of the triangle; algebraically ∂ = √3/2]
Build a square of side ∂,
and place the triangle vertically, riding its right side.
Together, these two figures take the entire width, with AD = ∂ + 1/2.

The diagonal of the rectangle type Δ, proportion √3, merges by definition with the side of the triangle. The J point is thus well on the diagonal of the rectangle.

The accounting is then simplified.

Any half equilateral triangle, right angle, has a small side 'a',
an hypotenuse '2a', and its vertical is a√3 (= 2a.∂).
The height of the figure is (∂ + 1/2) + 1,
or   ∂ + 3/2 = √3/2 + 3/2
The diagonal is 2 (∂ + 1/2) = 2∂ + 1 = √3 + 1

Subsidiary figure - 2

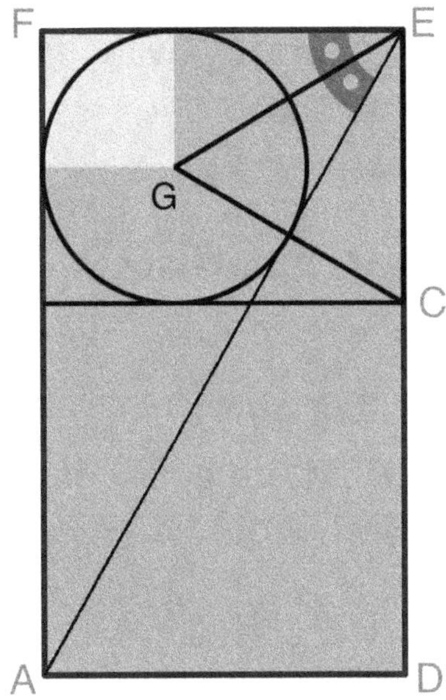

Fig. 81 : Subsidiary figure n° 2

The inscribed circle of a rectangle type Δ', is equal to the inscribed
circle of a triangle drawn by the diagonal of a rectangle type Δ.

A circle diameter 1 is drawn around the top, the point G,
of an equilateral triangle, side 1.
The diagonal EA of the rectangle AFDC is at 60° from FE
and G is on the bisector of the angle Â1 = FEA.
G is also on the bisector of the right angle in F.
It is equidistant from the FE and FA lines.
G is on the bisectors of two angles of the triangle FEA.
It is therefore the center of its inscribed circle.

The circle radius 1/2 is tangent to the three sides of the triangle FEA.
By definition, for the sides of the rectangle Δ' higher.
Then the diagonal AE is also the mediator of the triangle GEC.
It cuts the GC segment at right angle in the middle.
Thus the circle is tangent to AE at this point.

Symbolical resume

The basic square, inscribed, in the rectangle type Δ shows **by the difference** the side of the equilateral triangle of the rectangle type Δ'.

The inscribed circle of the triangle "half-rectangle Δ" gives us **directly** by its diameter, the side of the equilateral triangle.

Both mathematical facts explain why the figure of Conques didn't escaped to the Ancients (geometry with the eyes), as well as their fascination towards this proportion

4 – Monstrations "like Euclid"

The pedagogy of Euclid in the service of history

My colleagues of IREM propose demonstrations that I could compare to those of Euclid when he presents the Pythagorean theorem or the golden ratio. Euclid takes this opportunity to reveal the depths of geometry, especially the links that lead to algebra.

These developments are particularly valuable for the future of the conceptual validity of geometry with the eyes. In addition to the educational value of these exercises, it is likely that in this profusion of proposals we find the elements that will allow us to understand how men have gone from pure geometry to calculation.

The tablet Plimpton 322

For example, the tablet Plimpton 322, which sets a truncated series of Pythagorean triples. Our colleague Raphael Legoy has reconstructed the missing part of this listing, whereby a column of prime numbers appears (rather troubling for any "rational brain").

And this aspect is not necessarily the most interesting. The origin of this knowledge seems typical of the geometry with a grid. All the arguments that construct the series of Plimpton are perfectly accessible to the skills of geometry with the eyes. The tablet presents these triangles whole, from the most squared to the sharpest without forgetting any one. In this context, the prime numbers that appear by simple difference between two columns could reveal some secrets.

Monstration by Frédéric de Ligt

How to show that a rectangle type Δ is the sum of a square and a rectangle type Δ' ?

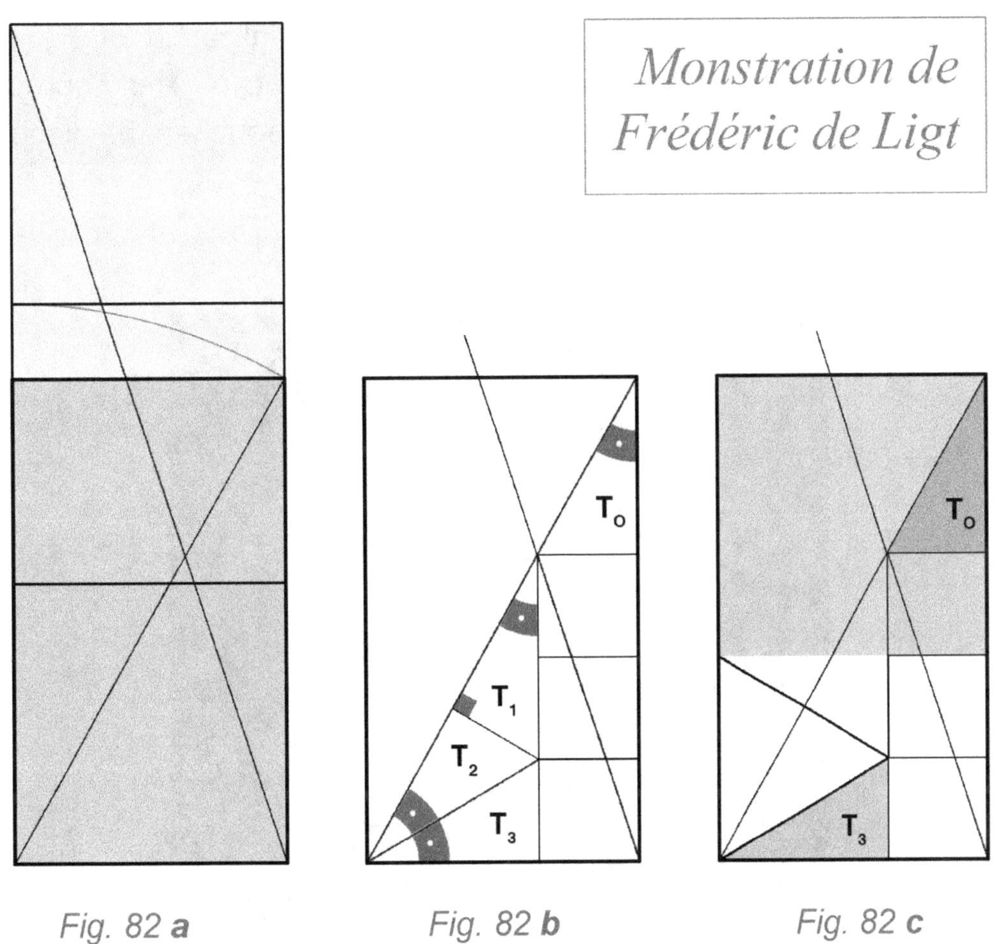

Fig. 82 a   Fig. 82 b   Fig. 82 c

Fig. 82 a – Inside a vertical triple square, we flap the vertical side of a double square - which becomes diagonal of a rectangle Δ. The two diagonals intersect at a point that allows us to build a triple square up to the bottom. The side of these squares will be the unit.

Fig. 82 b – Let T_1 with a 30° angle to the vertical, and an hypotenuse of two squares. T_1 is a half equilateral triangle. Its small side, at right angle to the long line, is therefore 1. T_1 and T_0 are isometric. For now, we know that the triangle T_2 which will search the angle at the bottom, has a small side of measure 1.

T_2 and T_3 form a kite together. Same sides of 1, and right angle. Now we know the sum of their angles below, 60°. The small angle of T_2 and T_3 is thus 30°, and therefore their hypotenuse is 2. T_2 and T_3 are isometric with T_1 and T_0, all halves of equilateral triangles of side 2.

Fig. 82 c – It is then easy to construct an equilateral triangle of side 2 resting on the side of the rectangle Δ. The width of the large rectangle Δ is thus [1 plus the height of the triangle]. Finally, a perfect square appears above this triangle and the double square.

Strong point of this monstration, the materialization of the triple square, which touches the diagonal of the rectangle Δ. This figure is to compare to that of the square side $\sqrt{3}/2$, which also touches the diagonal.

Decomposition of the hexagram

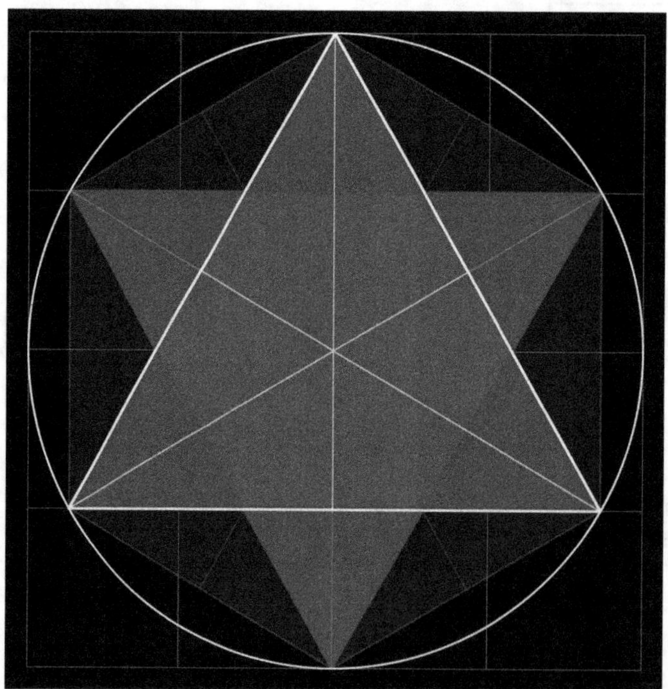

Fig. 83 : Decomposition of the hexagram

A generic figure shows these "half triangles". It is built in Egyptian style with a circle of diameter 4 on a 4x4 square.

At this occasion, we understand that the hexagram consists of two equilateral triangles that share their centers of gravity, of the inscribed and circumscribed circle, and their symmetry axis.

Monstration by Dominique Gaud

How to show that a rectangle type Δ is the sum of a square and a rectangle type Δ' ?

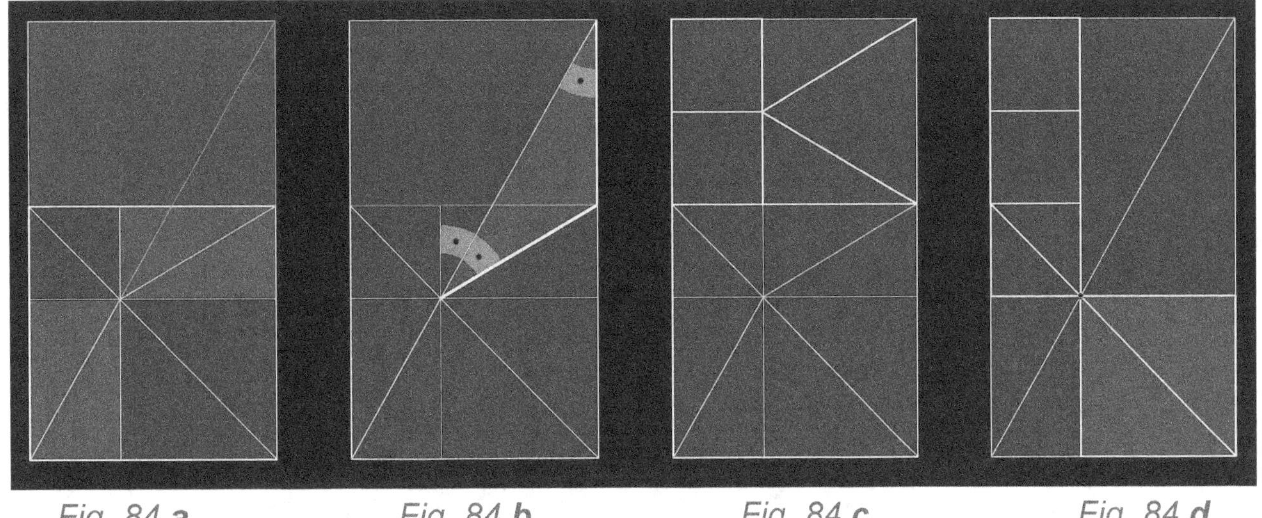

Fig. 84 *a*   Fig. 84 *b*   Fig. 84 *c*   Fig. 84 *d*

Fig. 84 a – Let a rectangle Δ, and its diagonal.
Let its inscribed square and its diagonal.
Let the vertical and horizontal lines at the crossing point of the diagonals.

The diagonal of the square (45°= 90°/2), is the symmetrical axis of the squares and the rectangles that we have drawn.
The diagonal of the two rectangles are symmetrical and the horizontal rectangle is Δ, since the vertical is Δ.

Fig. 84 b – The great angle of the diagonal of a rectangle Δ is 60°. The great diagonal is at 30°. Thus the angle between the little diagonal and the great is 30°.

Here the triangle, distinguished in light grey, is isosceles, because it has two angles equal to 30 °. The height of the residual rectangle is equal to the diagonal of the small rectangle.

Fig. 84c – To complete the figure of Conques, simply draw the diagonals of two superimposed rectangles Δ, and thus constitute an equilateral triangle (facing two squares of the same height).
The triple-square of the monstration of Frederic de Ligt is in watermark.

Fig. 84d – Strong point of this monstration

This demonstration shows the J point of the rectangle Δ, both as the corner of the square accorded to the height of the triangle, and as the corner of the triple square – already mentioned. This figure will allow to find another property of the rectangle type Δ .

Monstration of conclusion

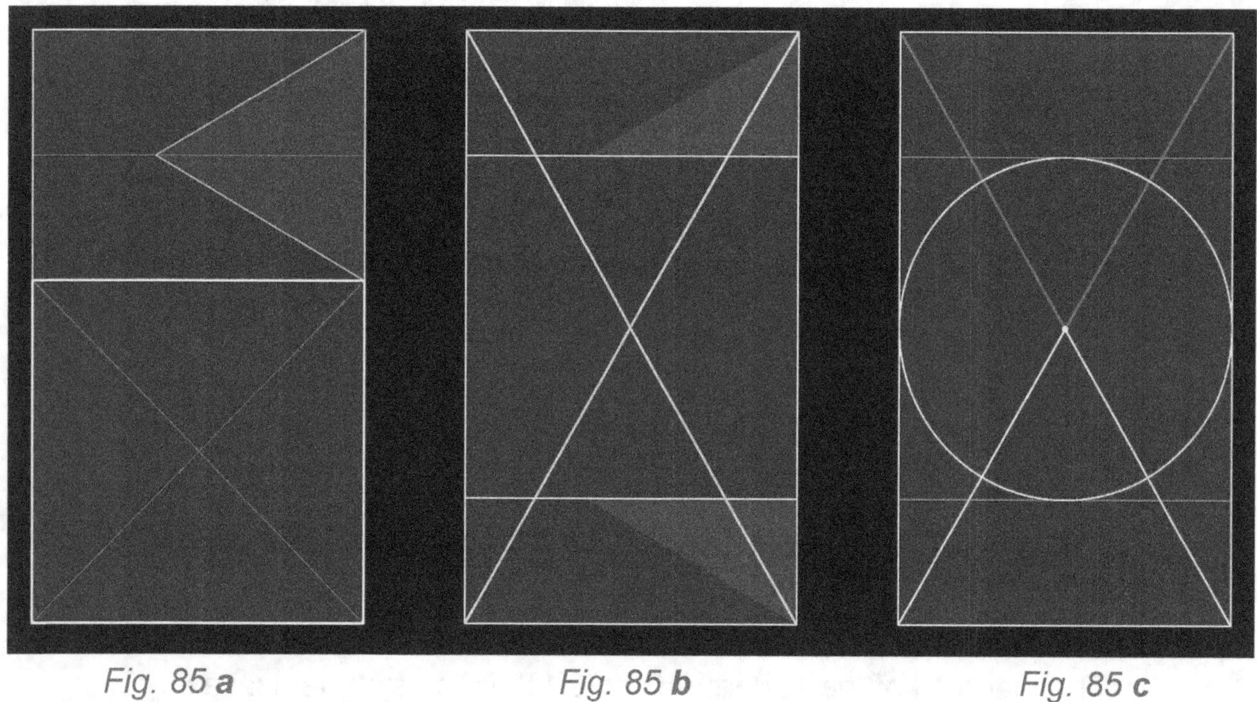

Fig. 85 *a*   Fig. 85 *b*   Fig. 85 *c*

What might have escaped...

Fig. 85 a – Consider the two bands above the inscribed square in the rectangle Δ. They each correspond to half of the triangle - height 1/2.

Fig. 85 b – Let's move one of these bands at the bottom of the rectangle Δ.

Fig. 85 c – We find again the triangle and circle. One of the definitions of the rectangle Δ'.

5 – Recapitulation

√3 and φ, the golden ratio

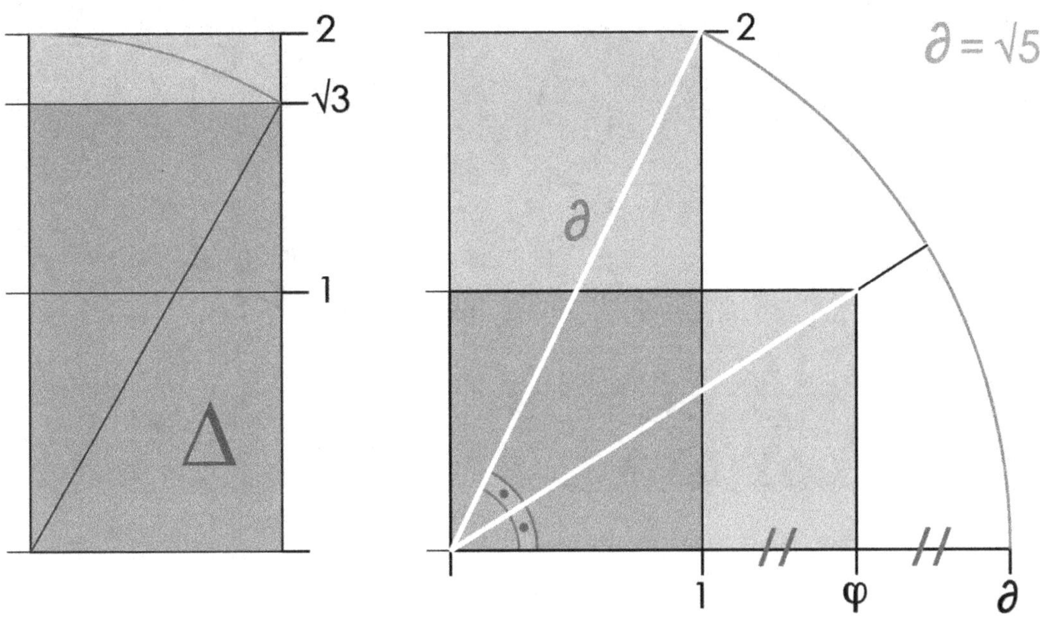

Fig. 86 (= Fig. 31) : Diagonals which produce √3 and φ

Left : The root of three is obtained by the copy of the two units of the double-square which become the diagonal of a rectangle, called type Δ. We construct indeed a right triangle with the measures 1 and √3, with 2 as hypotenuse. Indeed that's half of an equilateral triangle (hence Δ).

Right : The golden ratio is obtained by the half of the angle of the diagonal in a double square. The bisector intersects the horizontal of the first square to the distance φ. This is also the average of 1 and ∂ - diagonal of the double-square (=√5).

These two numbers come from the same double square by a game of diagonal. This common origin is the sign of a relationship. √3 confirms this exchange visually through the square.

The symbolism is the background of this "objective" study. All these structures are in the works of art for which they bring a meaning. And indeed, a geometric and algebraic properties bundle indicates that the root of three has a female nature in front of the golden ratio, masculine value. We can summarize this face-to-face by the Vesica Pisces of Venus and the pentagram of Mars.

The rectangles type Δ et Δ'

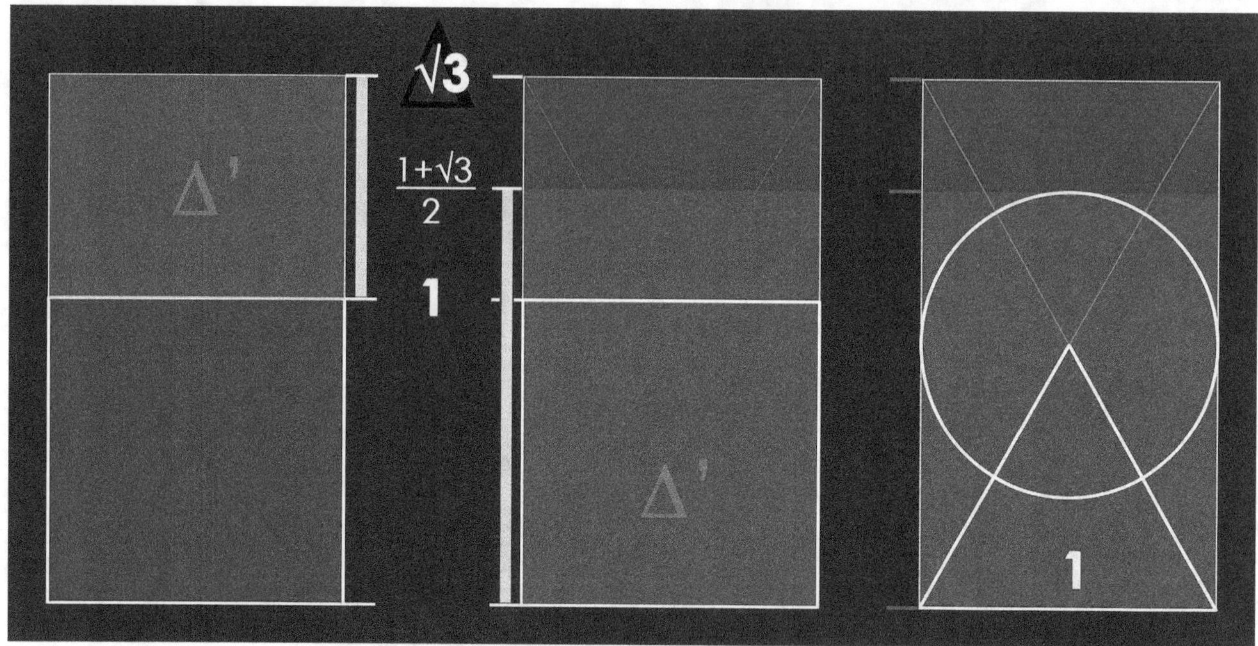

Fig. 87 : From geometry to algebra

Several formulas have their translation by geometry with the eyes like these two ways of seeing the proportion Δ'. We remove a square to a rectangle of type Δ, or we make the average of a square and a rectangle Δ.

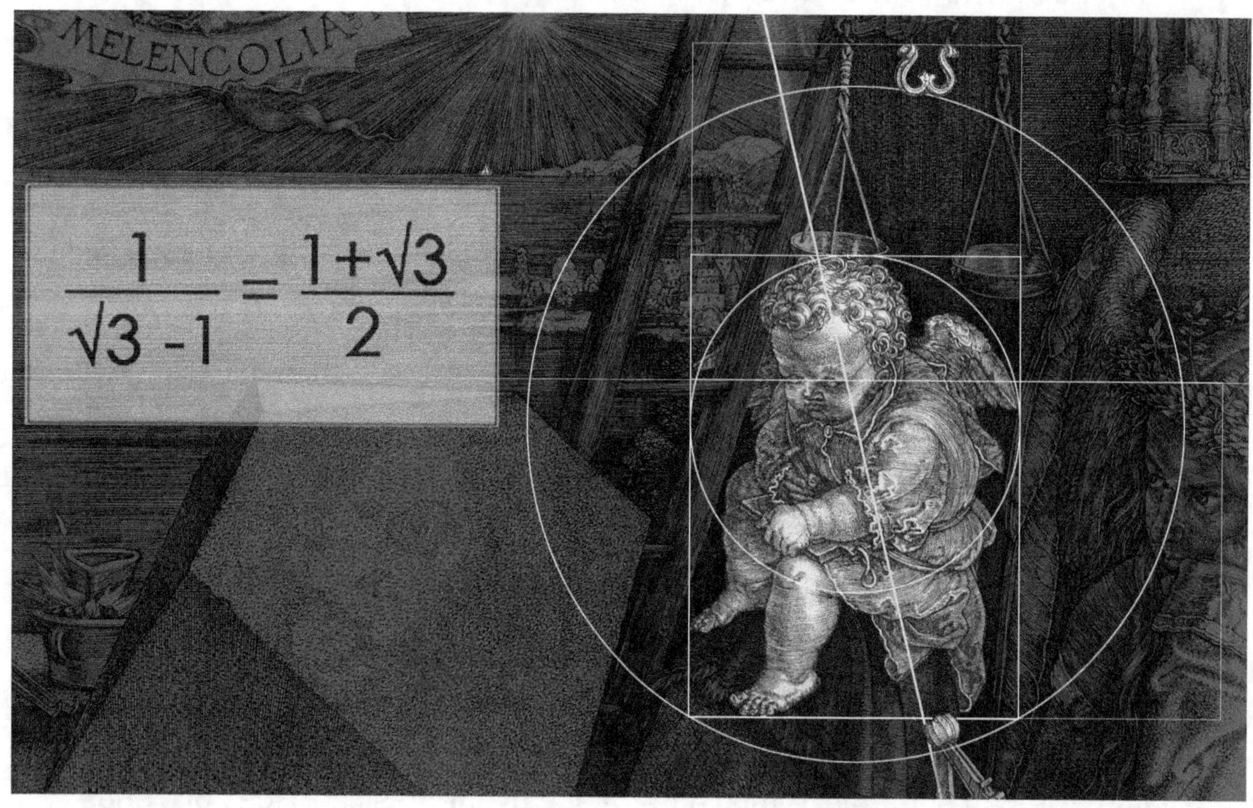

Fig. 88 : Cupid of Melencolia (A. Dürer)

IX – BABYLONIAN GEOMETRY

Without presuming trade that would have allowed the Egyptians to acquire elements of Sumerian and Babylonian mathematics, or their own ability to develop these elements by themselves, here are the developments about the triangle in the direction of calculation. This is the theoretical funds necessary to approach in particular a famous tablet, called Plimpton 322, and its lists of prime numbers.

1 – The trinity of angles in the triangle

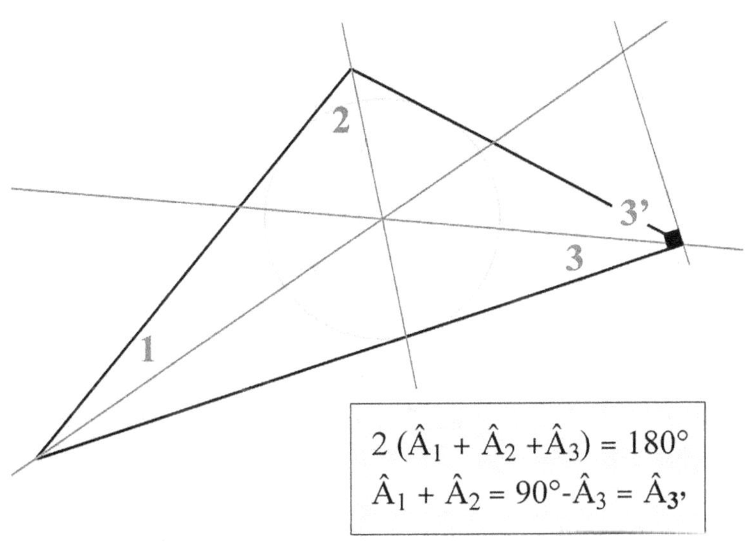

Fig. 89 : The sum of angles of a triangle

This property was inevitably known to the Egyptians. This is a direct consequence of the sum of the angles of the triangle.

The sum of the angles of two bisectors is equal to 90° minus the angle of the third :

$$2\,[\,\hat{A}_1 + \hat{A}_2 + \hat{A}_3\,] = 180°$$
$$\Longleftrightarrow$$
$$\hat{A}_1 + \hat{A}_2 = 90° - \hat{A}_3 = \hat{A}_{3'}$$

2 – The *paragonal* of right angled triangles

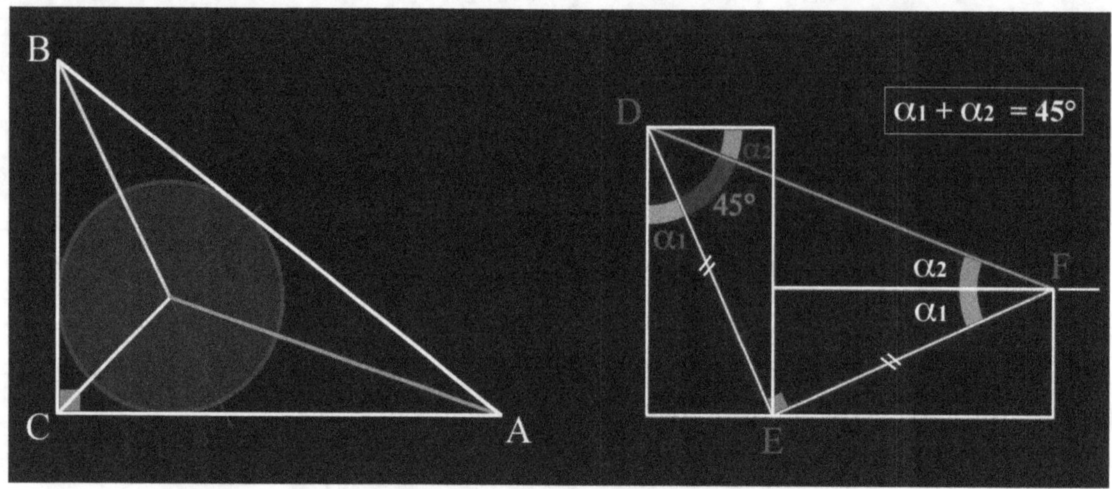

Fig. 90 : The "paragonal" of right angled triangles (DF)

Let a right triangle ABC, presented in its "classical position". We can consider that the bisector from B is the diagonal of a vertical rectangle. We note α1 the small angle of this bisector. We take the rectangle, called modulus, a second time but horizontally turned. Then we place it as an extension to the base.

An ensemble of two modulus shows the second bisector, from the first (except the one of the right angle). The right triangle can be summarized in two values. Its right angle allows to convert them "directly". However, instead of considering the apparent measures of the triangle, we enter into its internal structure.

A triangle DEF appears with a right angle, so two 45° angles =(180 – 90°)/2. But 45° is the angle of the bisector of the right angle in C of the initial triangle. It is also 90° – 45°, the result of the sum of the angles of the two other bisectors. I.e. the sum of α1 that we have defined, and α_2, the bisector of the angle in A. EF makes with the horizontal the exact angle of the bisector of the angle in A. Raphael Legoy has attributed the name of *paragonal* to this line.

The second bisector of a right-angled triangle is obtained by rotation of 45° of the first (the third concerning the right angle).

And $\alpha_1 + \alpha_2 = 45°$.

3 – The translation of the triangle into numbers

On their Neolithic grid, the Sumerians studied the most accessible right triangles, the smallest, and so they "have guessed the laws" : 3-4-5, 5-12-13, 7-24-25, 21-20-29 etc..

With these examples, the principle of monstration that we saw about the 3-4-5 triangle remains valid.

The hypotenuse of the right-angled triangle

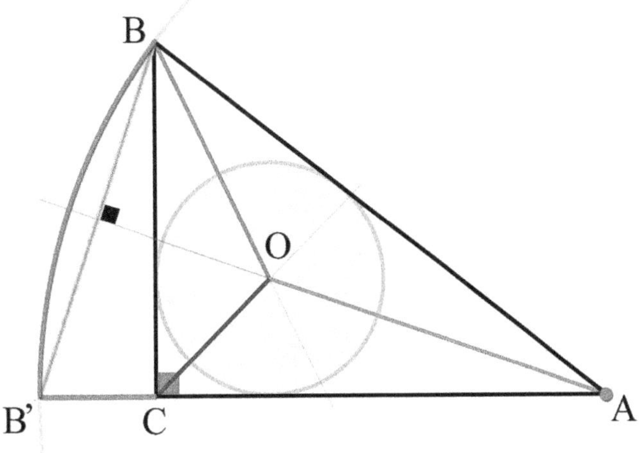

Fig. 91 : Compass deferral of hypotenuse

The first concern was to identify the grid points that are triangles with **integer sides**, particularly that of the hypotenuse. The Ancients have delayed the compass (the rope) this hypotenuse of one of the sides (called *cathetus*) of the triangle. And they tried all the rays that they were able to trace and they identified those who cross the grid points.

These examples are enough to allow speculate about the "laws" that govern their internal constitution. At this point, one speaks about rules, since the field is limited to the observable.

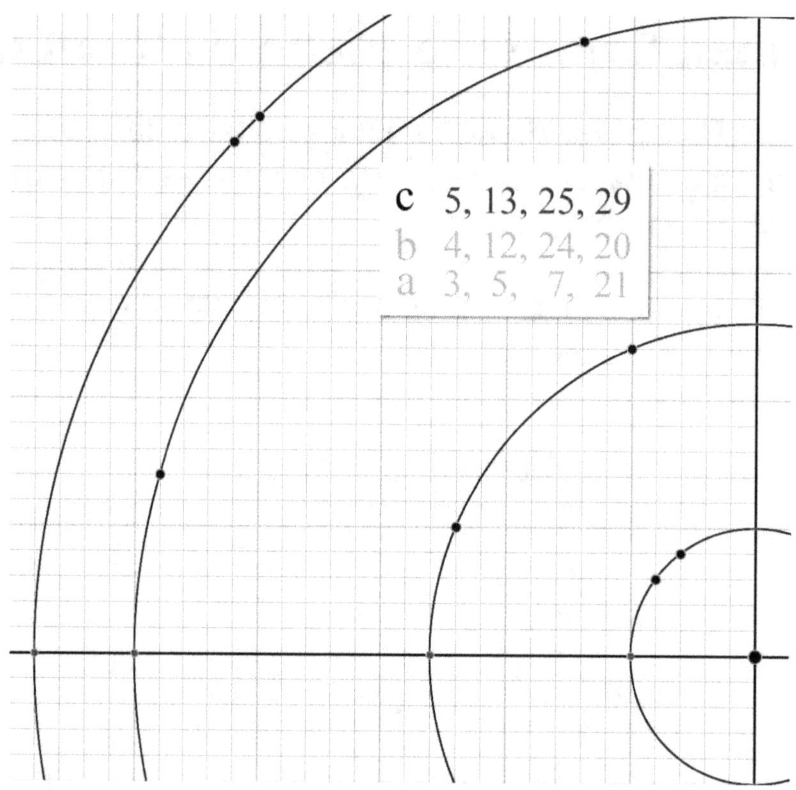

Fig. 92 : Compass deferral of the hypotenuse of integer triangles

The Neolithic geometry with the eyes is able to draw a small catalog of triangles with integer values.

The numerical laws of the right triangle

The example chosen here by Raphael Legoy (21-20-29 triangle) is complex enough to identify the laws of the triangle. The question is not only to identify the measurements of the triangle, or the relationship between the angles, but to define the laws that govern all (sides and bisectors).

1 -- This development requires writing (for calculation)

2 -- The reasoning is more complex than that which leads to the theorem of the diagonal (or Pythagoras).

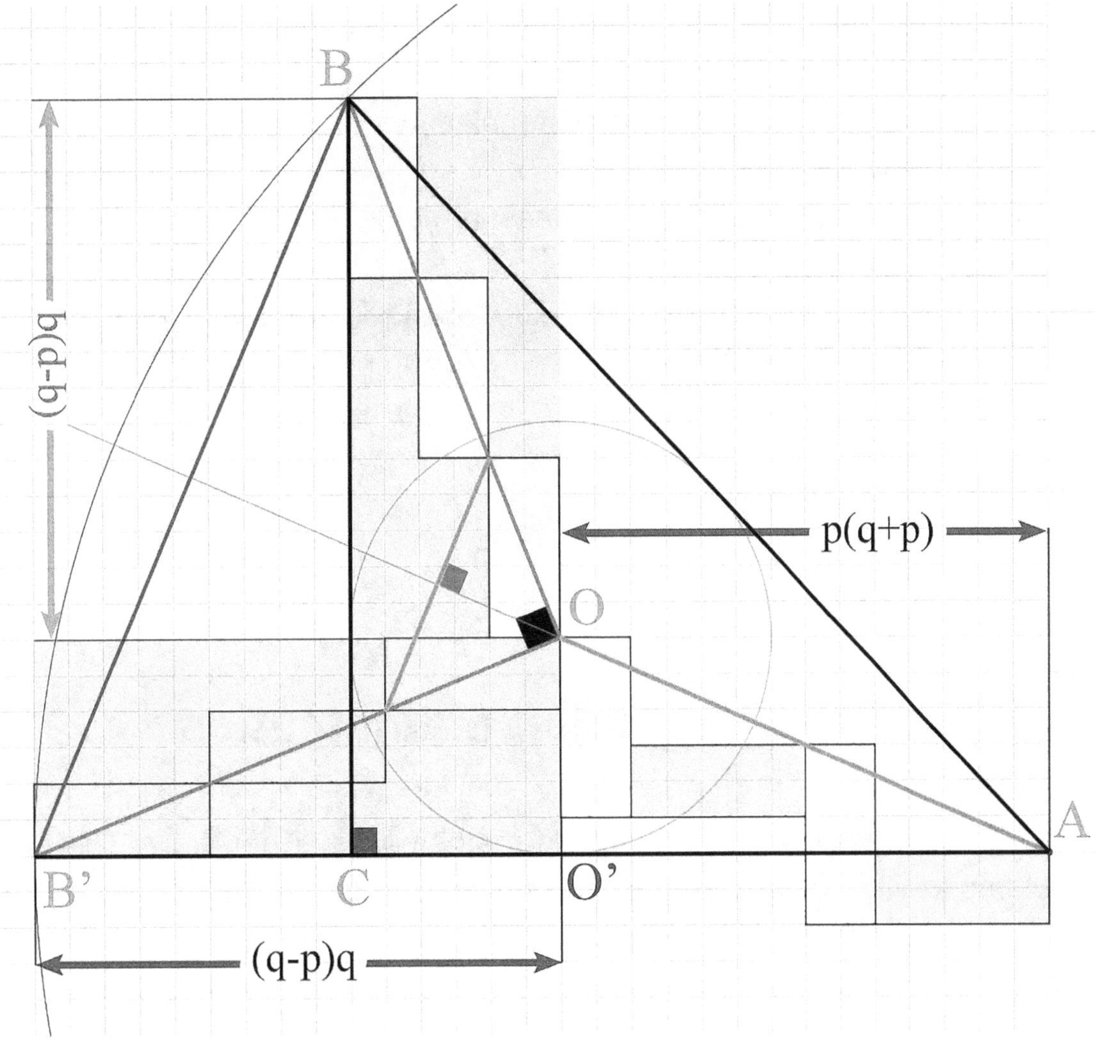

Fig. 93 : Study of the triangle 21-20-29

The figure of the *paragonal* is built here from a vertical modulus, width p and height q (white). This minimum rectangle (the pair (p, q) is irreducible and integers) accompanies the bisector coming from B.

We have seen that the bisector coming from the vertex B of the triangle is automatically linked to that one which goes to A, on the right (figure of *paragonal*). Then the point B ', symmetric to B with respect to this *paragonal* builds a segment B'O which is right angle to the bisector too, BO. The proof is on the picture : the diagonals of two grey and white modulus just to the left of O are right angle (and their *paragonal* is orthogonal to OA).

This figure allows assumptions that other triangles will confirm.

The ratio of p and q here is 5/2. It is irreducible.

Note that three modulus are required to cover OB.

        Now $3 = q-p$

        We also note that two assembled modulus cover OA

        Now $2 = p$

From these assumptions, the measurements of the triangle are simple calculation.

$$BA = B'O' + O'A = \quad (q-p)q + p(q+p) \quad = q^2 + p^2$$
$$CA = CO' + O'A = \quad (q-p)p + p(q+p) \quad = 2pq$$
$$CB = \quad\quad\quad\quad\quad\quad (q-p)p + (q-p)q \quad = q^2 + p^2$$

What manuals relate as :

$$a = q^2 - p^2 \qquad b = 2pq \qquad c = q^2 + p^2$$

4 – The level of Babylonian mathematics

The assessment of Babylonian mathematics is not easy. The two problems of the study are, firstly, the over-interpretation of data, for example by applying the apocryphal processes (Euclidean) on results that will not cease to be fair, while they are the result of another approach.

Then, we must admit that certain reasonings of the Babylonians are literally "in breach with ours", without deserving the anathema. The notion of subjectivity can even be inversely involved to what it intends to denounce: our definition of objectivity can paradoxically be translated into a kind of bias that is detrimental to the historical truth of this culture.

X – HISTORICAL PANORAMA

1 – The History of Mathematics

The origin, development and propagation of this particular geometry in history are not the primary concerns of this research. However, a few arguments arise which can become a contribution. For example, without recourse to calculation - and thus before writing - these mathematics respect the rules that the Greeks will precise with their "hypothetico-deductive system". The logical spirit of the Greeks is in fact the vast didactic development of a knowledge which they inherited from the Mesopotamians and the Egyptians.

Conversely, the beginnings of geometry, in the Upper Paleolithic, do not integrate with this stream. Signs of this period, described as proto-writing, cannot be considered as writing in the strict sense. In both cases, the explicit structures are lacking to organize the signs and figures. It has long been assumed that geometry had appeared in the Paleolithic, faced with the virtuosity prowess of the *Magdalenians* in their stone and bone industry. The progress of archaeological research and the reflection they inspire seem to identify a very abrupt border between the populations of nomadic hunter-gatherers and those who settle around agriculture.

This change in lifestyle is accompanied by a change of thought, especially spiritual. We go from shamanism to religion as we commonly hear it today and, above all, we move from magic to sacred. Thus the wall of the Paleolithic caves is not yet a plan: it is the membrane that separates the world from here and beyond. In particular, Olivier Keller's work can be cited in this regard. (see bibliography)

Similarly, the site of Catalan Huyuk does not present the level of geometrical organization of the Eanna belonging to the civilization of Uruk IV. Is it any wonder that geometry is involved ?

2 – Egypt – Written sources

The exact role of legendary Egypt is being debated. We can not avoid this question because for a long time in this research, the geometry with the eyes was accredited to the Egyptians, and for good reason : the greatest master of this art, Albrecht Dürer, refers to Egypt in at least two of his artworks. He placed pyramids on a horizon where they have seemingly functional role.

("Adoration of the Trinity", 1511 and "MELENCOLIA § I", 1514).

We have very few documents written by the Egyptians because papyrus (ex Rhind, Moscow etc.) is an extremely fragile material - unlike the clay of the Babylonian tablets. Moreover, these documents are exercises of scribes when according to legend, the knowledge of the priests is hidden. The scarcity of vestiges could explain this situation by itself, but we have indirect sources from Greek and Latin authors. They make explicit reference to Egyptian sacred mathematics, in terms that give a reality to the legend.

Greeks in Egypt

- Thales (-625, -547) He studies geometry in Egypt.
- Pythagoras (-580, -497) He learns the language in Pharaonic Egypt then "the secret doctrines of Gods".
- Democritus (-460, -370) This philosopher learns geometry in Egypt.
- Eudoxus (-406, -355) This astronomer spends 15 months with Egyptian priests.
- Plato (-428, -347) After the meeting with the Pythagoreans, he stays in Egypt with the priests of the high clergy.

Src. 1 — Plutarch (~ 46, ~ 125)

Original Greek, major thinker of Rome, precursor of neoplatonism.
French text : PLUTARQUE OEUVRES MORALES
TOME V — TRAITÉ D'ISIS ET D'OSIRIS — § [56]
Trad. française : V. BÉTOLAUD & D. RICHARD
http://remacle.org/bloodwolf/historiens/Plutarque/isisetosirisa.htm

Summary : *According to Plutarch, the divine perfection of nature comprises three principles : the understanding, the matter, and the product of their combination - the world. He takes up the philosophy of Plato with the idea / father, the mother / matter and the begotten.*

The triangle 3-4-5 from Egypt represents these principles, and the sum of the small squares is equal to the great. The side 4, horizontal and feminine, is that of Isis. The side 3, vertical and masculine, is that of Osiris. The hypotenuse (5) is the side of their son Horus, result of the operation of the one and the other. Plutarch then throws himself into an explanation of the numbers and etymology of names, which he puts in parallel. In the same way he never ceases to see in Plato's discourse the echo of that of the Egyptians. / Summary

We discover the existence of a 3-4-5 triangle, Egyptian according to its gods, which none of the documents known to Egypt mention. The mathematics of papyrus deal only with calculation when those which Plutarch exposes are religious and without calculation. The scribes and the priests do not have the same preoccupations.

We also discover the transmission of the "knowledge" of the priests. Should we consider it as esoteric, theological, religious, metaphysical, philosophical or symbolic ?

Src. 2 — The Presocratics

« Les Présocratiques », Collectif, Biblio. de la Pléiade, n° 345, 1988
http://www.la-pleiade.fr/Catalogue/GALLIMARD/Bibliotheque-de-la-Pleiade/Les-Presocratiques
TRANSLATION FROM FRENCH TEXT – THE MILESIANS – XI

Proclus.

Just as the exact knowledge of numbers was born among the Phoenicians as a result of trade and business, likewise was it among the Egyptians that geometry was invented, for the reason I have stated. Thales was the first Greek to bring back this matter for speculation to Egypt; he himself enriched it with numerous discoveries, and bequeathed to his successors the principles of many others, going farther in abstract generalization and sometimes in empirical investigation.

De même que la connaissance exacte des nombres prit naissance chez les Phéniciens du fait des échanges commerciaux et des affaires, de même est-ce chez les Égyptiens que fut, pour la raison que j'ai dite, inventée la géométrie. Thalès fut le premier Grec à rapporter d'Égypte cette matière à spéculation; lui-même l'enrichit de nombreuses découvertes, et légua à ses successeurs les principes de nombreuses autres en allant plus loin tantôt dans la généralisation abstraite, tantôt dans l'investigation empirique. (Commentaire sur le premier livre des Éléments d'Euclide, 65,3.)

Plutarch

As claimed, Thales and Hippocrates of Chios, the mathematician, have made trade; And Plato covered his traveling expenses by selling (olive) oil in Egypt (Solon, 2).

It is believed, because this is have been learned from the Egyptians, that Homer and Thales posed water as the principle of the origin of all things (Isis et Osiris, 34)

Thalès, à ce qu'on prétend, et Hippocrate de Chio, le mathématicien, ont fait du commerce; et Platon couvrit ses frais de voyage en vendant de l'huile en Égypte (Solon, 2).   … C'est, pense-t-on pour l'avoir appris des Égyptiens, qu'Homère et Thalès posaient l'eau comme principe de l'origine de toutes choses (Isis et Osiris, 34)

Flavius Josephus

Everyone agrees that the firsts among the Greeks to have studied celestial and divine things, as Pherecydes of Syros, Pythagoras, and Thales, were the pupils of the Egyptians and the Chaldeans, and left few writings; These writings pass in the eyes of the Greeks to be of all the oldest, and scarcely do they believe even that they have really written them. (Against Apion, I, 2.)

Tout le monde s'accorde à reconnaître que les premiers chez les Grecs à avoir étudié les choses célestes et divines, comme Phérécyde de Syros, Pythagore et Thalès, furent les élèves des Égyptiens et des Chaldéens et laissèrent peu d'écrits; ces écrits passent aux yeux des Grecs pour être de tous les plus anciens, et à peine croient-ils encore qu'ils les aient vraiment écrits. (Contre Apion, I, 2.)

Aetius

He studied philosophy in Egypt and returned to Millet, who was already very old. (Opinions, I, III, I.)

Il étudia la philosophie en Égypte et revint à Millet déjà fort âgé. (Opinions, I, III, I.)

Iamblichus

Thales advised Pythagoras to go to Egypt, and to converse as often as possible with the priests of Memphis and Diospolis. It was from them that he had derived all this knowledge, which made him appear as a wise and learned man in the eyes of the crowd. (Pythagorean Life, 12)

Thalès conseilla à Pythagore de se rendre en Égypte et de s'entretenir le plus souvent possible avec les prêtres de Memphis et de Diospolis : c'est d'eux qu'il avait tiré toutes ces connaissances qui le font passer pour sage et savant aux yeux de la foule. (Vie pythagorique, 12.)

Herodotus

It is [in Egypt], in my opinion, that geometry was invented, and it was from there that it came to Greece. (The Inquiry, II, 109.)

C'est [en Égypte], à mon avis, que le géométrie fut inventée, et c'est de là qu'elle vint en Grèce. (l'Enquête, II, 109.)

Himerius

Pindar has sang the glory of Hiero in an Olympic ode, and Anacreon the fortune of Polycrates, while the Samians sent their gifts to the goddess. Alcaeus praised Thales in his odes, when Lesbos celebrated a panegyric. (Selected pieces, 30, Codex of Naples, Col. XXVIII, 2)

Pindare a chanté la gloire de Hiéron dans une ode olympique, et Anacréon la fortune de Polycrate, alors que les Samiens envoyèrent leurs présents à la déesse. Alcée fit l'éloge de Thalès dans ses odes, lorsque Lesbos <célébra> une panégyrie. (Morceaux choisis, 30, Codex de Naples, col. XXVIII, 2)

Pliny

Thales of Miletus has found a method of measuring the height [of the pyramids] by measuring their shadow at a time when it is regularly equal to its object. (Natural History, XXXVI, 82.)

Thalès de Milet a trouvé une méthode pour mesurer la hauteur [des pyramides], en mesurant leur ombre à l'heure où elle est régulièrement égale à son objet. (Histoire naturelle, XXXVI, 82.)

Plutarch   (p. 22)

Simply putting a stick at the end of the shadow of the pyramid and making two triangles with the line made by the Sun's ray, touching the two extremities, you showed that there was such a proportion of the height of the pyramid with that of the stick, like there is the length of the shadow of one with the shadow of the other. (The Banquet of the Seven Wise Men, 2, p.147 A.)

Dressant simplement un bâton au bout de l'ombre de la pyramide, et se faisant deux triangles avec la ligne que fait le rayon du Soleil en touchant aux deux extrémités, tu montras qu'il y avait telle proportion de la hauteur de la pyramide à celle du bâton, comme il y a de la longueur de l'ombre de l'un à l'ombre de l'autre. (Le Banquet des sept sages, 2, p.147 A.)

/ EXTRACT

Thus the Greece of philosophers and mathematicians inherited from Egypt priests - and artists. This allows us to specify the cultural space where geometry with the eyes has found a place to develop, and to be transmitted. In a second step, the question of the origin of this "practice" arises. We then turn to Mesopotamia, without ignoring more distant spaces like India and China. Knowledge and ideas circulate as surely as goods and people, and a geometry without writing constitutes an ideal platform for exchange. This principle will apply later between the Muslim and Christian worlds.

3 – The compositions as sources

1 — The Pyramids of Giza

The composition has the reputation of being an oral tradition. On the study side, geometry with the eyes comes straight from the works and the plans. The ground plan of the pyramids of Giza appears to be appropriate : it is deployed like geometry on a flat surface. The study at : http://www.contemporary-painting.com/Yvo_Jacquier-3_squares_Giza.pdf

Extract : (A golden gift for the study) The plan of reference is the work of the physician John A.R. Legon. He gathered all the data of well-known egyptologists into a synthesis. Especially M. Lehner, 1985 - J.H Cole,1925 - Flinders Petrie, 1880. This plan has been approved by he Archaeology Society of Staten Island.
http://www.john-legon.co.uk/gizeplan.htm

This is a real gift for the study. The comparison of the geometry to the reality of a work is the most delicate stage of the study. Above all, this step is the most "questionable". / Extract

In the particular case of Giza, we confront geometry with attested measures, with very narow margins - not to say extreme - that the composition will all respect. The interpretation of the geometric motifs completes the study. The Giza plateau presents itself as a family story in which Queen Henutsen, mother of Khéphren and wife of Cheops, plays a major role despite the administrative name of her pyramid (G1C). The figures also refer to the Egyptian religion, notably the star of the pharaoh which joins the sky of Nout. The composition of the Giza plan uses all the vocabulary that geometry with the eyes can know. The golden ratio, is displayed with four digits after the decimal point.

2 — The Egyptian frescoes

Some paintings also have a geometry of composition. However the pictorial techniques of the Egyptians do not have the precision of those they demonstrate in architecture and sculpture. Perhaps it is a question of material, to which is added the wear and tear of time. It should be noted that many of these achievements are bas-reliefs.

Fresco of Benia (Pahekamen ou Paheqamen)
Tomb TT 343, wall east, part south, 18e dynasty, circa 1500 BC :
http://www.contemporary-painting.com/comparative-geometry/B02-Benia.jpg

The golden ratio is displayed here very clearly, according to a rectangle that measures 6 patterns in width - at the very top. The characteristic golden divisions intervene at the elbow of the character twice. This would not be the case if, for example, the motif were part of a general rectangle of 8/5.

The spatial arrangement of the hieroglyphs seems to be governed by geometry. This other field of research is as vast as that of the Renaissance which for the moment occupies the study.

4 – The character of Pythagoras

Pythagoras (– 580 to – 497, 6th century BC) is clearly not the discoverer of that which symbolizes his theorem. Vedic India or ancient China practice long before him an areas geometry, based on what is called the *theorem of the diagonal*. Moreover, in the famous tablet Plimpton 322, Babylonians show prime numbers at the confines of very sophisticated calculations about the right angled triangle.

Cultural exchanges have always existed, especially at a time when borders were not subject of conflicts (for example, in the 4th millennium BC). It is more than probable that Southern Europe has captured the echoes of what was happening in the East.

However, we can not treat Pythagoras as a usurper. The complexity of the character is in the image of the Pythagoreans his disciples, and the disciples of his disciples. The "subject" of Pythagoras is as fascinating as that of the golden ratio, and the historical aura of the master tends to capture anything that approaches the spheres of his vast knowledge. He marks the entry of the Greeks into an era of prosperity, whose theorem has become the symbol. This world of logos will (re)define the rules in all things, to the point that it still serves us as a reference today. Pedagogy and democracy are the perfect examples.

Pythagoras is on the path to Euclid, pure mathematics - detached from their utility. At the same time he makes experiments on music and harmony. He still teaches a mystical and qualified knowledge of esoteric. The question that arises in this book concerns its contribution to geometry with the eyes, as the basis of composition in the arts.

There is no great danger in claiming that the Pythagoreans have done a thorough study of the meaning of mathematics, which is summarized in the master's formula "all is number". The interpretation of structures is based on numbers. However, there is nothing to prove that this work is subsequently fully integrated with that which he went to seek in Egypt on the advice of Thales. We do not know the influence of *Pythagoreanism* on art, on the construction of cathedrals and Renaissance painting.

5 – The geometry of composition

Blaise Pascal denounces two excesses in his "Thoughts" :
To exclude reason and to admit only reason.

Sacred geometry does not take geometric forms for simple objects : it uses structures whose definition to be complete does not stop at that of the figures. Numbers accompany construction, they are indispensable for translation into human language. In the end, without the numbers the geometry remains silent and without the forms the numbers are blind. It is necessary to associate and to understand both aspects in the approach of translation.

A knowledge is not limited to its method; it pursues goals and, by necessity, it is accompanied by a moral charter (science shows us today the problems this entails). The knowledge of composition developed in several distinct cultures - particularly in Egypt and Mesopotamia - also in China and India. When it comes to art and architecture, we can not detach works from the motivation of their authors.

Thus geometry with the eyes is not defined solely by a coherence which would bypass the theorem of the diagonal or of Pythagoras. This geometry has a meaning in all the civilizations that have practiced it, from that of Uruk to the Western Renaissance. It is used to translate spiritual values through lines that underlie works of art and architecture. Sacred geometry intends to bear the meaning of a work as much as the physical work.

It should be noted that the perspective system is incapable of playing such a role. It makes it possible to represent reality in a plausible way, in its strictly material aspect, yet its lines bear no particular value. For this reason, artists of the Renaissance practiced simultaneously the two systems of composition within the same works, assuming in this way all the preoccupations of the representation.

By the purity of its art, Egypt justifies a popularity rating. In the future, it is likely that other civilizations such as Babylon and Uruk will receive comparable attention.

6 – Area and Writing

The theorem of Pythagoras is the didactic expression of that of the diagonal. Here we have a fine example of the progressive path of mathematics. The final result is accepted by everyone, from the geometer of the sacred to the accountant of the treasure, including the Euclidian pedagogue. And this unanimity has the effect of erasing all the history !

Certain figures of geometry with the eyes foreshadow the demonstration of the theorem. Jean-Paul Guichard warns us : it is possible to construct schemes that are more influenced by Euclid's discourse than by the concerns of "geometers of the time". Clearly, the monstrations are potentially apocryphal because they can feed on the experience of the Greeks and the algebraists ! This book can help to clarify the status of figures in that some are indispensable to the building of the corpus, when others respond to didactic needs today.

Then a reality must be taken into account: Pythagoras' theorem demands writing. The concept of area and calculation, which are associated with it, are impracticable without writing. The historical background makes it possible to give a reality to this idea. Indeed, writing is born in the Neolithic era, when these parameters are explicitly united: well beyond a simple productive sedentarization, we are witnessing the regrouping of populations in large cities, struck with a social and urban organizational chart . Two principles emerge then, tax and property, and both require the practice of writing and the concept of area (as quantity). Without calculation of area and without tablets to note who paid what and to whom, the Neolithic cities would have been unmanageable.

The first writings are, moreover, books of account. At the same time, geometry with the eyes will retain the tradition of non-writing. The grid on which everything is understandable, demonstrated and retained, will remain the mark of this culture.

Now we must sift through the geometric constructions that seem to prefigure Pythagoras, and identify those that have allowed the passage to written thought. This is the inflection point of history. The figures below are somehow lures. On the other hand, in the simmering of the concept of area, they will emerge.

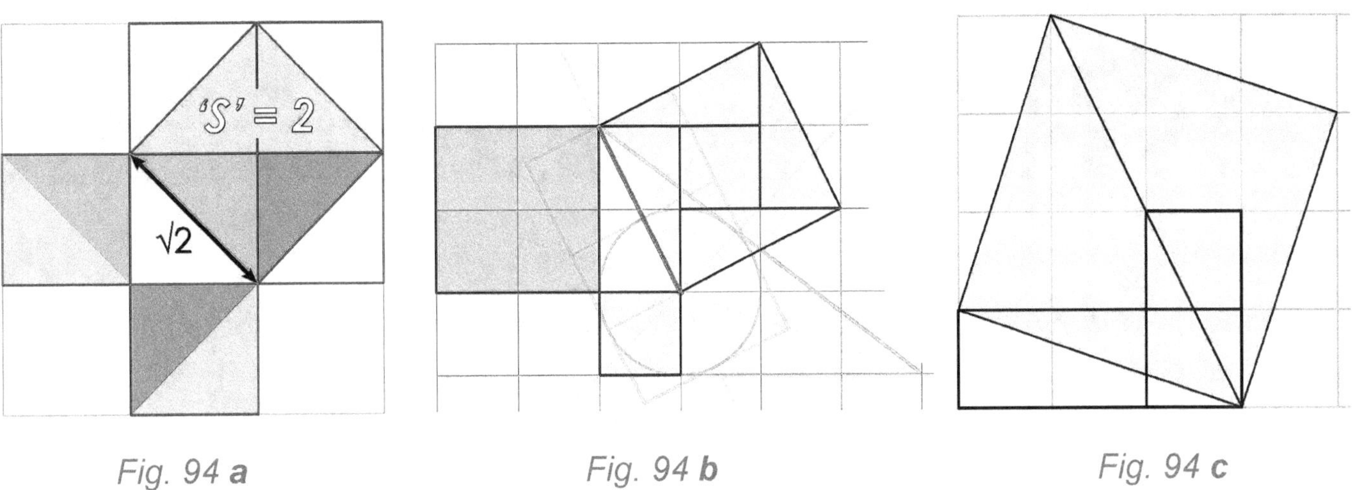

Fig. 94 *a*   Fig. 94 *b*   Fig. 94 *c*

N.B. : √2 and √5 are not defined under this form by the geometry with the eyes. In contrast, in a didactic way, the areas of 2 and 5 squares are here explicit.

Fig. 94 c – This figure 5 enunciates a very different principle :

The angle between the diagonal of an horizontal triple-square and that of a vertical double-square is the angle of the diagonal of a simple-square.

There is no question of area; there is no more need to write.
This is (still) geometry with the eyes.

This figure is essential to the corpus. The result is involved in several monstrations (see Fig. 49) and more importantly, it summarizes the relationship between the angles of the three bisectors of the triangle 3-4-5. This is a real basic. Then, it is enough to pose the definition of the area to solve the diagonal of the triple square. This figure strangely reminds that which serves to demonstrate the theorem of the diagonal.

Fig. 95 a – Jean-Paul Guichard reports indeed the most direct demonstration of the Pythagorean theorem. Simply juxtapose a square of side a at the foot of a square of side b (with b> a). And then we remark that the two rectangles marked with a cross have the same area = ab

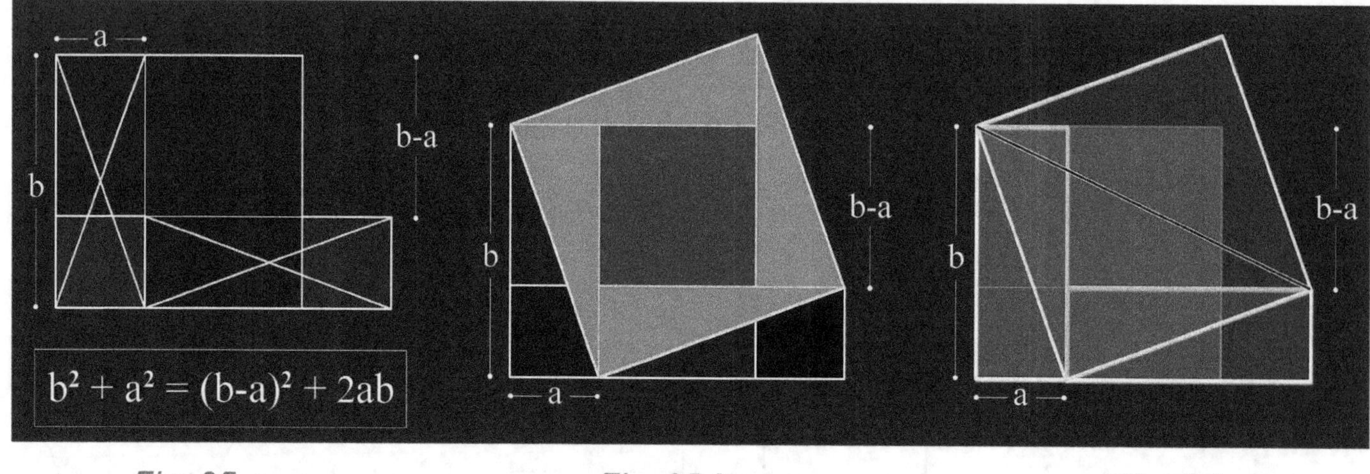

Fig. 95 *a*   Fig. 95 *b*   Fig. 95 *c*

Fig. 95 a – The area of the large square of side b plus the small one of side a, is equivalent to the sum of the empty square, $(b-a)^2$, with the two rectangles marked by their diagonals.

Fig. 95 b – The way of generalization of visual evidence is in this figure 10. It is found in ancient China, long before Pythagoras. This figure could be also the proof of the theorem of the diagonal in the *Sulbasutras* (see Olivier Keller). It can finally be considered as known by the Babylonians and Egyptians. We find it back in the developments of Clairaut, that stands out completely from the Euclidean approach in his Elements of geometry (in 1741), where he presents the proof of the theorem of Pythagoras as a natural proof that anyone can find, in looking for where is the base point from which are drawn the 2 "diagonals".
(J-P Guichard)

Fig. 95 c – The figure of the *paragonal*, generalization of Fig. 94, does not need the concept of area in its demonstration. The logic of angles is enough. Both arguments by forms (angles) and numbers (area), lead to the same figure. And the transition from one conception to another corresponds to the emergence (simultaneous ?) of the concepts of area and writing.

7 – Two examples of pedagogy

Jean-Paul Guichard explains : if we complete the half-square to get a square, the figures become demonstrations of the Pythagorean theorem.

First approach

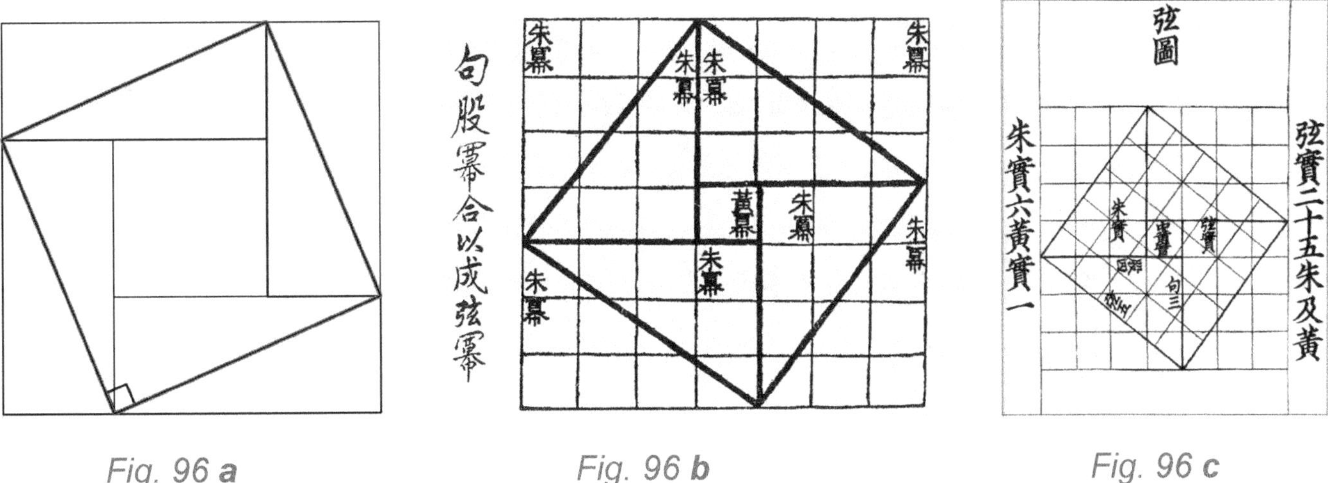

Fig. 96 a   Fig. 96 b   Fig. 96 c

Fig. 96 a – Again, the two rectangles and their diagonals are rotated a quarter turn and we obtain Fig. 96 a) which is a classic of Chinese mathematics to show the theorem of Pythagoras. It reads easily on a grid (Fig. 96 b). The illustration is made for the sacred triangle !

Fig. 96 b – This figure is called in Chinese « Figure of the hypotenuse ». *The Zhou Bi Suan Jing*, or else *Chou Pei Suan Ching* is one of the oldest Chinese mathematical texts. The title literally means "The classical arithmetic of gnomon and the circular paths of heaven". This book dates back to the Zhou Dynasty (1046-256 BCE), but its compilation including adding of materials continued up to the Han Dynasty (206-220 AD).

Fig. 96 c – We find the same figure in an edition of « Ten classic calculation » dating of 1213, a work that compiles a series of ancient texts, made during the Tang Dynasty (618-907 AD).

Second approach

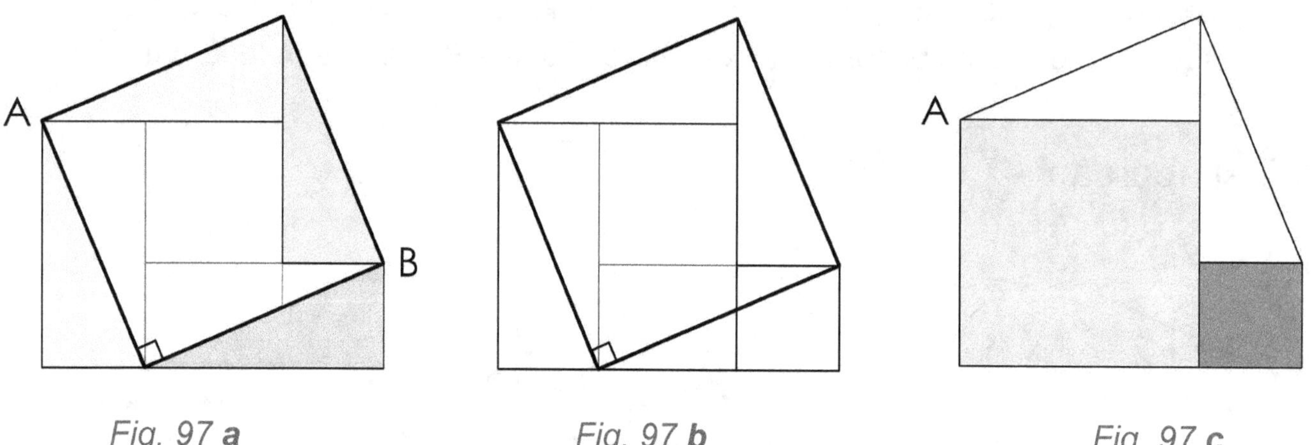

*Fig. 97 **a***   *Fig. 97 **b***   *Fig. 97 **c***

We make turn the two triangles (dark grey and light gey, Fig. 97 a) of a quarter turn around the points A and B (the cons) and we get Figure 97 b.

This is the figure used by Clairaut in his « Elements of Geometry » (1741), to prove the Pythagoras theorem (Fig. 97 c).

XI – THE INTERPRETATION

1 – The choice of words

With a perfect mastery of the subject, the title of this chapter would be "translation of geometry". But we have not reached that stage yet, this adventure has just begun. In the current state of studies, it is legitimate to confess our interpretation. This is not because it prolongs the original discourse of geometers, but because it tries to straddle the abyss that we still have to fill.

This work reconstructs a corpus of properties as fantastic as they are unknown. This is the solid part of the study of composition. In the works, the observation of the figures must respect margins of precision. This purely geometrical discourse is confronted with the observation of works - down to the smallest detail. Finally, every language demonstrates recurrent structures, and the comparison of works makes it possible to highlight them. One term summarizes this methodology : "comparative geometry".

The venture is not new : to translate the geometry and numbers of the symbolic, to identify their meaning. The term "Tradition" refers to a set of knowledge where, if there is no evidence and consistent coherence, there is a certain consensus. One can not ignore the existent, nor content oneself with denouncing its contradictions. Some of this knowledge is fair and must be identified.

2 – The meaning of 3 and 4 in question

According to the said "tradition", the number 3 is celestial, feminine and sacred, facing the number 4, terrestrial and masculine. During my studies I sought and found many arguments that go in this direction. However, if the proofs of the 3 Celestial are direct, those of the 3 feminine are indirect and subjective.

In addition, the Vesica Piscis, the vulva *deic* and feminine, symbol of Venus, is marked by the √3, and not directly by the 3. At the same time, Saturn's case is troubling : its Byzantine magic square is of order 3*. The planet Saturn marks the limit of the visible - with a naked eye, in this the option of the 3 Celestial is legitimate. On the other hand, the virility of the *old taciturn* does not fit with the feminine gender.

> **Ref. *** : *A magic square is a set of consecutive integers, starting from 1 (1, 2, 3, 4, ... n^2), and arranged in a square: as many rows as columns. The term "order" denotes the number of cells on one side (n). This square is "magic" if all the sums of the numbers are the same, regardless of the row and column. The Byzantines assigned, according to orders   3 – Saturn, 4 – Jupiter, 5 – Mars, 6 – Sun, 7 – Venus, 8 – Mercury, 9 – Moon.*
>
> *During the Renaissance, Luca Pacioli transmitted the mathematical knowledge of the Byzantines to the Western Europe, including that of the magic squares. Trithème and Agrippa are the main authors who will deepen their symbolic meaning.*

3 – What cultures are we talking about ?

The stream of civilizations that constitute the field of study will doubtless have to find a generic name. For the moment, we are content to follow the trace of a geometry with the eyes that emerges in Mesopotamia before the advent of writing, and which is transmitted for five millennia until the Renaissance - its apotheosis.

Historical background

The geometry with the eyes is Neolithic. There is no Paleolithic trace of its first figure: the Vesica Piscis.
(Information taken from Genevieve von Petzinger)

Second half of the Fourth Millennium BC.
This geometry is found in the floor plan of the great temple, the Eanna, dedicated to the goddess Inanna / Ishtar.
2,500 years BC
The floor plan of the Giza Pyramids is a demonstration.

The Greece of Phidias (5th century BC) and Pythagoras (6th century BC) deepened the theoretical principles.

Rome is on the path, yet the Romans are more engineers than artists.

Alexandria is the crossroads of all cultures

Byzantium remained for more than a thousand years the great school of sacred geometry; The icons become major patterns.

The iconoclasm (8th and 9th centuries) led this culture to exile headed for the first time towards the West, up to Ireland

(Book of Kells). Charlemagne protects this culture.

Geometry accompanies the orthodox expansion from the 9th C. in the direction of the East (Moravia, then Bulgaria, then Kiev)

After the iconoclasm, Byzantium continues to be the center. Its monks are the only ones in Europe to read Greek in the integral.

The fall of Constantinople, in 1453, provoke a second exile - Venice is the main destination.

The Renaissance corresponds to the apotheosis of this art. The tarot, as an encyclopedia of symbolism, is the testament of this multi-millennial culture. Dürer is its last master.

Afterwards: the works of certain painters (as J.D. Ingres, who stayed in Italy) include elements of this practice.

Other civilizations

The approaches can be very different depending on the continents. What about India, China, the Americas ? Others will certainly answer. We will take an example closer to us by exchanges, and for which we have a minimum of elements.

The West African tribes do not practice the geometry exposed in this book. On the other hand, the philosopher Lucien Stéphan showed the mastery of dynamic principles in African statuary. A geometry not theorized by the artists who is none the less in their will. Numbers also make sense in these micro-cultures inside their large puzzle. They are not directly related to geometry, as is the case with "Western" sacred geometry. (see bibliography - Stéphan)

This may be one of the reasons for the disparity of conceptions between societies - particularly about 3 and 4. Another probable cause is the nature of the symbolic reference. In the case of the West, this frame of reference is spiritual and sacred; The symbolism of the lineal (lineage) societies studied by Sylvie Fainzang plays above all a socially inegalitarian role. The anthropologist exposes the variations of the conception of 3 and 4 in these words :

« If it works like that in many societies in West Africa, particularly in the Bisa of Burkina Faso, the reverse system is found in many other countries, particularly among the Mano, Kpelle, Sherbro, Kissi, Kono and Bété of Sierra Leone and Liberia, as well as among the Serer and Mandingo, where the 4 is on the contrary attributed to the man and the 3 to the woman. » (see bibliography – Fainzang)

4 – The Celestial, the Terrestrial and the Gender

According to the logic of the Byzantine magic squares - the reference of so many authors, the confrontation of the Celestial with the Earthial is reflected in a dialogue between Saturn and Jupiter. This is precisely the theme of the most famous esoteric work of all time : « MELENCOLIA § I », engraved by Albrecht Dürer in 1514. The characters on the stage are two angels and their identity is now revealed : Saint Michael And Cupid. Dürer's message is precise : the question of gender about 3 and 4 is that of the sex of angels. It is therefore elsewhere that we must find arguments. On the other hand, one of Melencolia's many lessons is that Saturn reigns over time while Jupiter dominates space. Note: Greek philosophy comprises three forms of time: Chronos, Aiôn and Kairos. (see bibliography - Y Jacquier et G Beys-Salvan)

The 3 Celestial and the 4 Terrestrial must be considered neutral. These values are sufficiently "generic" not to be overloaded by other notions, especially those of the gender.

The 7 Catholic virtues are a good example of this Celestial / Terrestrial sharing. They are distinguished in 4 cardinal (terrestrial) virtues and 3 theological virtues (celestial). Strength, justice, prudence and temperance are of the order of education. Faith, hope, and charity are divine graces. It should be noted that all are feminine without expressing the feminine gender. This logic is found again in the divine trinity (celestial) and the four gospels (testimonies of the life of Christ on Earth).

The triangle of Isis (3-4-5)

One can cite the Egyptian triangle (of Isis), which associates the 3 to Osiris, the 4 to Isis, and the 5 to Horus. In its horizontal position, it is logical to consider the 4 as earthly - and not as feminine. The 3 vertical expresses the Celestial; This is in accordance with the myth of Osiris - which rules over the afterlife. The 5 sloping of Horus would simply not make sense in front of a definition of the genres for the 3 and the 4.

It should be noted that the "tradition" of which we have spoken does not explain the inversion of its genres in the Egyptian symbolism. The mystery is not enough to escape this question. Does this mean that symbolism is a matter of conventions ? In the symbolism that we study, numbers intersect geometry. The ancients thus constructed a system whose logic rests on mathematical properties.

We can then understand the unity of this culture of composition. It has gone through periods, civilizations, peoples and countries without fundamentally changing its conception of forms and numbers. Very early, men understood the geometry that we reconstitute, enough not to have to "change option" afterwards.

5 – The sacred feminine and the masculine

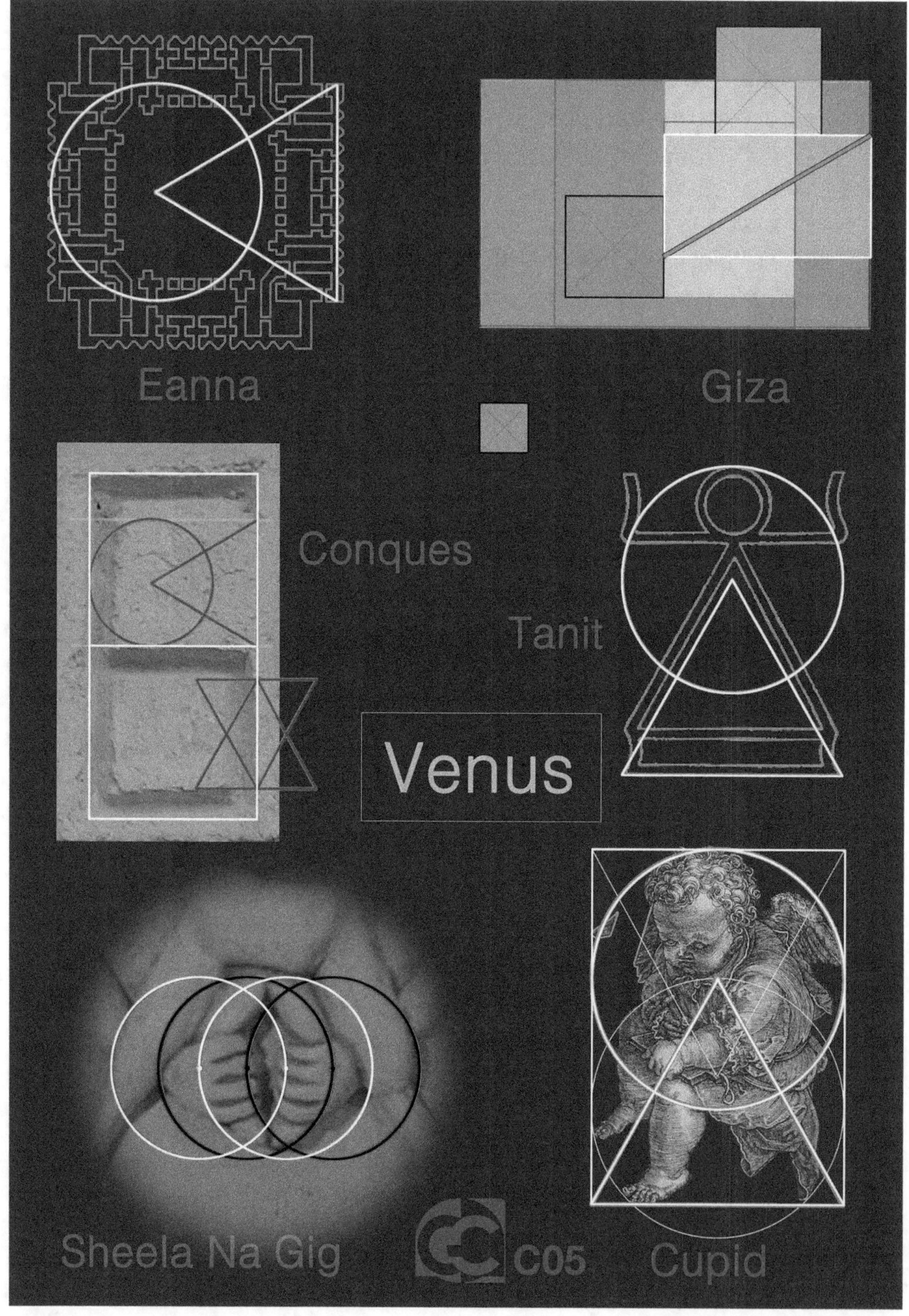

Fig. 98 : Venusian figures through the time

The geometric figures representing Venus, Aphrodite, Inanna / Ishtar, Tanit etc belong in the majority of cases to the "family Δ". These forms are governed by the √3, the hexagram is their generic and the Vesica Piscis their first expression.

It should be emphasized that the mandorles of Christ are never strict Vesica Piscis. The center of a circle is always offset from the twin circle where it should land. Other geometric rules, different according to the case, decide their arrangement.

Originally, the Venusian cults associate femininity and fertility. Fertility (ability to engender) is itself confused with fecundity (capacity for abundance). And their benevolence extends to every form of life, especially to agricultural production.

Faced with the hexagram and its √3 – family called Δ, the pentagram represents the masculine : it is the movement facing the matrix. The number of gold (φ) enters its constitution, and the golden spiral is a good example to evoke movement.

Fig. 99 : Composition of "The Birth of Venus" by S. Botticelli – 1485

Fig. 99 – This layer of composition of the painting - there are several, includes the main subjects of this work : a Vesica Piscis; Two rectangles 3x4 (double triangles 3-4-5) inclined according to the logic of the pentagram. Finally a golden spiral emerging from the bottom of the sky (kingdom of Uranus) to fertilize the foam. Chloris the nymph runs with great strides along this path, in quest of eternal beauty.

The head of the beauty is outside the almond, respecting the definition of the fish body. To top it all off, the shape of a tail makes Venus a mermaid, echoing the original title of the painting: "Venus Anadyomene", that is, going out of the waters. It should be noted that the conception of Venus reminds strangely that of fish.

6 – The tarot encyclopedia

The tarots of Marseilles, that which Nicolas Conver perpetuates from the model in 1760, were designed by Albrecht Dürer.

This project of Byzantine origin found refuge in northern Italy, at the fall of Constantinople. The painters' guilds took up the challenge, but they needed an engraver of genius to ensure the sustainability of the project. The master hand that completed the cards was that of Albrecht Dürer. Four unusual engravings accompany the cards - the Meisterstiche. The queen of the four, Melencolia, serves as a backdrop to the presentation of the tarot blades.

The tarot cards are physically (in reality) compatible with Melencolia. There is room for three cards in width on the engraving. Through transparency, symbols exchange their secrets. The Ternary VII-VIII-IX of the Chariot, Justice and Hermit is particularly eloquent. The horizon cuts the cards in half. The rainbow makes a triumph to the character of the Chariot and the weight scales answer to each other, united by the letter omega. This scales of the last judgment belong to the archangel (St Michael), below, and it symbolizes the Parousia, or return of Christ (the alpha corresponds to its birth, and the omega to its return). The astrological sign of the Libra is itself represented by an Ω, and this detail appears on the tympanum of Conques. Finally, the Hermit (representation of Saturn) grabs the hourglass of Melencolia and the lantern of the cards. He eats the magic square of Jupiter. To respect the legend, this one is in stone.

Fig. 100 : Blades VII-VIII and IX of the tarot cards on Melencolia

In the background of these drawings, the composition of the tarots and of the engraving are built on one same grid. After five centuries, the precision of the engraving, in its respect of the grid, is still of the order of a tenth of a millimeter.

What Was to be Shown

APPENDIX – TRIGONOMETRY

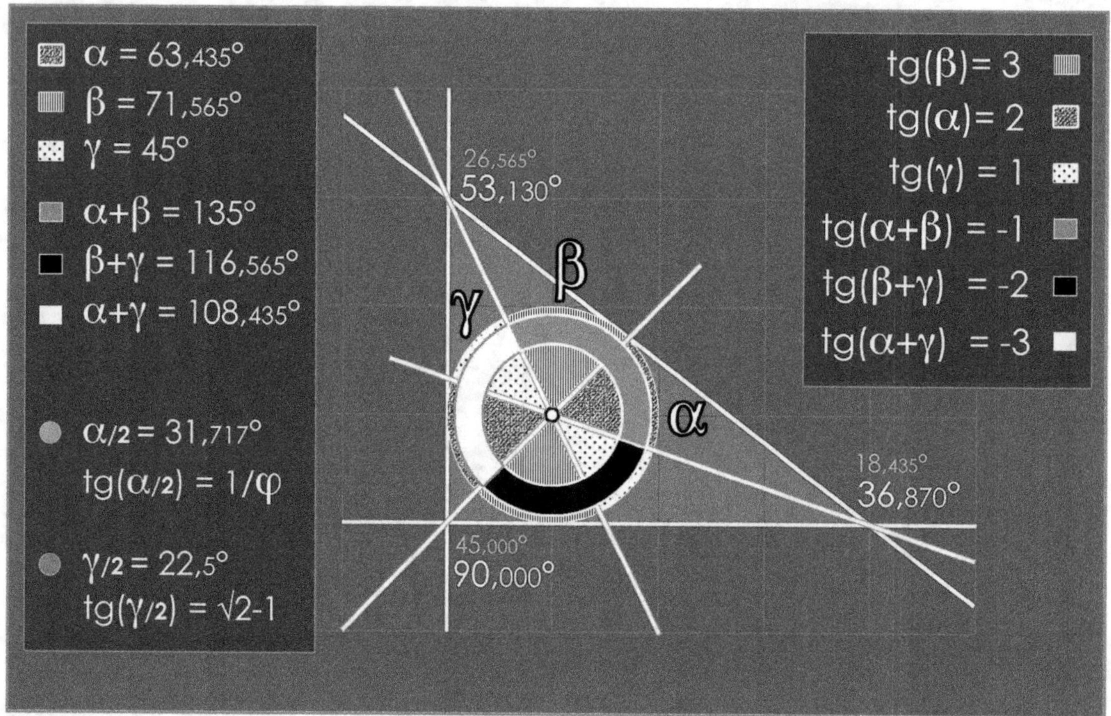

Fig. 101 : Trigonometry of the triangle 3-4-5

The bisectors which intersect at the center of the incircle defines naturally three angles. Anthony Rédou has been the first to notice the free values exhibited by certain central angles of the triangle : 135° and 45°... And their tangents are equally impressive ! Trigonometry is not antique (it develops really at Renaissance). It nevertheless reveals the internal organization of the 3-4-5 triangle.

The order of a bisector is the number of squares of which it is naturally the diagonal. Two by two, bisectors have two angles : one is acute and the other is obtuse
- The tangent of the acute angle is "the order" of the third bisector.
- The tangent of the obtuse angle is "the order" of the third, preceded by the sign "-".

BIBLIOGRAPHY

PREHISTOIRY

The Mind in the Cave CONSCIOUSNESS AND THE ORIGINS OF ART – Lewis-Williams, David, Thames & Hudson Ltd, London, 2002 – ISBN for USA and Canada: 978-0-500-77044-3 – UK and the Rest of the World: 978-0-500-77030-6

Aux origines de la géométrie – Olivier Keller, Vuibert, 2004, http://independent.academia.edu/OlivierKeller

L'Invention du nombre, Des mythes de création aux Éléments d'Euclide – Olivier Keller, Classiques Garnier, 2016 – ISBN 978-2-406-05971-4

Elements for a prehistory of geometry – Olivier Keller (Irem de Lyon, France), Communication to the HPM conference, Uppsala 2004

BABYLON & EGYPT

Unexpected LINKS between EGYPTIAN and BABYLONIAN Mathematics – Joran Friberg, World Scientific, 2005 – ISBN 981-256-328-8

Count Like an Egyptian – Reimer, David, Princeton University Press, 2014
ISBN 978-0-691-16012-2 (hardcover : alk. paper)

Mathematics in Ancient Egypt A Contextual History – Annette Imhausen, Princeton University Press, 2016 – ISBN 978-0-691-11713-3 (hardcover : alk. paper)

ARCHITECTURE AND MATHEMATICS IN ANCIENT EGYPT – Corinna Rossi, Cambridge University Press, 2003 ISBN 13 978-0-521-82954-0 hardback 13 978-0-521-69053-9 paperback

Cultural Memory and Early Civilization Writing, Remembrance, and Political Imagination – Jan Assmann, Cambridge University Press, 2011 – ISBN 978-0-521-76381-3 Hardback ISBN 978-0-521-18802-9 Paperback

Ancient Egyptian Science A Source Book – Volume Three : Ancient Egyptian Mathematics – Marshall Clagett, American Philosophical Society, 1999 – ISBN 0-87169-232-5

Ägypten Theologie und Frömmigkeit einer frühen Hochkultur – Jan Assmann, Kohlhammer Urban- Taschenbücher, 1984 – ISBN 3-17-008371-6

GREECE

Heaven and Earth in Ancient Greek Cosmology From Thales to Heraclides Ponticus – Dirk L. Couprie, Springer ScienceBusiness Media, 2011 – ISBN 978-1-4419-8115-8 e-978-1-4419-8116-5

Thales von Milet in der frühen christlichen Literatur – Andreas Schwab, De Gruyter, 1977 ISBN 978-3-11-024598-1 (hardcover : alk. Paper) 978-3-11-024599-8 (e-ISBN)

Die Milesier: Thales Herausgegeben von Georg Wöhrle – Mit einem Beitrag von Gotthard Strohmaier, Walter de Gruyter GmbH & Co, 2009 – ISBN 978-3-11-019669-6

The Pythagorean Plato And The Golden Section; A Study In Abductive Inference – Scott Anthony Olsen, Dissertation Degree Of Doctor Of Philosophy – University Of Florida, 1985

THEMATIC

Sacred Geometry Philosophy and Practice – Robert Lawlor, Thames & Hudson Ltd, London, 1982, 2002 – ISBN 0-500-81030-3

Le matin des mathématiciens – Collectif dont M. Caveing et E. Noel, Ed. Belin, Paris, France, 1985 – ISBN 978-2701105338

Introduction à l'Origine de la Géométrie de Husserl – Jacques Derrida, PUF, 1962
ISBN 978-2-13-057916-8

CONNECTIONS The Geometric Bridge Between Art and Science – Jay Kappraff, World Scientific Publishing, 2001 – ISBN 981-02-4585-8 ISBN 981-02-4586-6 (pbk)

La sculpture africaine Essai d'esthétique comparée – J Kerchache, J-L Paudrat, L Stéphan, G Viatte, Editions Citadelles et Mazenod,1988-2008 – ISBN 978 2 85 088 441 2

Les Sexes et leurs nombres
Sens et fonction du 3 et du 4 dans la société burkinabé – Sylvie Fainzang, Revue L'Homme, Année 1985, Volume 25, Numéro 96

« MELENCOLIA § I » d'A. DÜRER Battre les cartes des réalités
Conférence de Y Jacquier et G Beys-Salvan, Colloque de Meknès, 23-24 mars 2016
http://www.jacquier.org/meknes/JACQUIER-SALVAN-2016.pdf

ROMAN & GREEK PHILOSOPHY

The Presocratic Philosophers A Critical History with a Selection of Texts
G. S. Kirk, J. E. Raven, M. Schofield, Cambridge University Press, 1964, 1983

The Presocratic Philosophers – Jonathan Barnes -- Vol. I.Thalesto Zeno – Vol. 2, Empedocles to Democritus --The Arguments of the Philosophers – London : Routledge & Kegan Paul, 1979.

Handbook of Greek Philosophy From Thales to the Stoics Analysis and Fragments –
Nikolaos Bakalis, Trafford Publishing, 2005

Les Présocratiques LES MILÉSIENS – XI – Collectif, Biblio. de la Pléiade, n° 345, 1988
http://www.la-pleiade.fr/Catalogue/GALLIMARD/Bibliotheque-de-la-Pleiade/Les-Presocratiques

Moralia Volume V – Treaty of Isis and Osiris – Plutarch

PLUTARQUE – OEUVRES MORALES - TOME V — TRAITÉ D'ISIS ET D'OSIRIS — **§ [56]**
http://remacle.org/bloodwolf/historiens/Plutarque/isisetosirisa.htm

DIDACTIC

Les angles sous tous les angles – Tangente Hors-série N° 53, Pôle Paris, 2014

Enseigner les mathématiques au collège à partir des grandeurs en sixième :
IREM de Poitiers, 2010 — les aires — les longueurs
http://irem2.univ-poitiers.fr/portail/index.php?option=com_content&view=article&id=110:enseigner-les-mathematiques-en-sixieme-a-partir-des-grandeurs-les-longueurs3&catid=11&Itemid=123

Les constructions mathématiques avec des instruments et des gestes
Inter-IREM collectif, coordonné par Evelyne Barbin, Ellipses, 2014
http://www.editions-ellipses.fr/product_info.php?products_id=10020

THANKS

Historical and Pedagogical Adviser

Jean-Paul Guichard - IREM de Poitiers.

Partners, Friends and Researchers

Sébastien Bassu – Strasbourg
Guillaume Beys-Salvan – Rennes
Christophe de Cène – Combourg
Thierry Ciblac – Paris
Sylvie Fainzang – Paris
Thibaut Gress – Paris
Olivier Keller – Paris
Zdeněk Halas – Prague
Henri Lombardi – Besançon
Alain Lambert – Annonay
Frédéric de Ligt – Poitiers
Raphaël Legoy – Le Havre
Hanna Lakoma – Prague
Jan Makovsky – Prague
Jean-Paul Mercier – Poitiers
Yvan Monka – Strasbourg
Isabelle Paresys – Lille
Geneviève von Petzinger – Victoria, CAN
Eurydice Prentoulis – Bonn
Anthony Rédou – Prague
Pierre Théon – Saint-Malo

Websites

http://www.contemporary-painting.com — Website of the artist
http://www.jacquier.org — French version
http://www.melencoliai.org — Albrecht Dürer
http://www.art-renaissance.net — Renaissance
http://www.andrei-rublev.com — Icons

Table des matières